Contributions to Management Science

More information about this series at http://www.springer.com/series/1505

Ayşe Özmen

Robust Optimization of Spline Models and Complex Regulatory Networks

Theory, Methods and Applications

Springer

Ayşe Özmen
Ankara, Turkey

ISSN 1431-1941 ISSN 2197-716X (electronic)
Contributions to Management Science
ISBN 978-3-319-30799-2 ISBN 978-3-319-30800-5 (eBook)
DOI 10.1007/978-3-319-30800-5

Library of Congress Control Number: 2016939943

Printed on acid-free paper

This Springer imprint is published by Springer Nature
The registered company is Springer International Publishing AG Switzerland

To my parents Mehmet Zeki and Ayşin, my sisters Özlem and Pınar, and my brother Ali

Contents

Acronyms

AM	Additive Model
ANN	Artificial Neural Network
AAE	Average Absolute Error
BFs	Basis Functions
CART	Classification and Regression Tree
CIs	Confidence Intervals
CD	Conic Dual Problem
CGPLM	Conic Generalized Partial Linear Model
CMARS	Conic Multivariate Adaptive Regression Splines
CP	Conic Primal Problem
CQP	Conic Quadratic Programming
r	Correlation Coefficient
CV	Cross-Validation
dim	Dimension
DR	Dynamic Regression
GAMs	Generalized Additive Models
GCV	Generalized Cross-Validation
GLMs	Generalized Linear Models
GPLM	Generalized Partial Linear Model
int	Interior
IPMs	Interior Point Methods
LS	Least-Squares
LSE	Least-Squares Estimation
LP	Linear Programming
LR	Linear Regression
MSE	Mean-Square Error
MARS	Multivariate Adaptive Regression Spline
MLR	Multivariate Linear Regression
PRSS	Penalized Residual Sum of Squares
RSS	Residual Sum of Squares
RCGPLM	Robust Conic Generalized Partial Linear Model

RCMARS	Robust Conic Multivariate Adaptive Regression Splines
RCQP	Robust Conic Quadratic Programming
RGPLM	Robust Generalized Partial Linear Model
RMARS	Robust Multivariate Adaptive Regression Splines
RO	Robust Optimization
RMSE	Root Mean Squared Error
SOCP	Second-Order Cone Programming
SDP	Semidefinite Programming
SVD	Singular Value Decomposition
TE	Target-Environment
TR	Tikhonov Regularization
WRCMARS	Weak Robust Conic Multivariate Adaptive Regression Splines
WRMARS	Weak Robust Multivariate Adaptive Regression Splines

Chapter 1
Introduction

Multivariate adaptive regression splines (MARS) [46] is a form of non-parametric regression analysis for building high-dimensional and nonlinear multivariate functions and applied in many fields of science, engineering, technology, finance and control design in recent years. It is a modern methodology of statistical learning, data mining and mathematical estimation theory which is important in both regression and classification, and develops an multiplicative-additive model in a two-stage process, namely, forward and backward, without specific assumptions about the underlying functional relationship between the variables [58, 68]. Continuing on the success of MARS in modeling real-life problems, as an alternative to MARS, Conic MARS (CMARS) [127, 136] was developed for the backward part of the MARS algorithm in a previous study. For this approach, a Penalized Residual Sum of Squares (PRSS) is employed for MARS as a Tikhonov regularization (TR) problem [5], and then, it is treated with a continuous optimization technique, namely, Conic Quadratic Programming (CQP) [12].

For both the MARS and CMARS models, however, data are assumed to contain fixed input variables whereas, in reality, the data involved in regression problems contain noise. Therefore, these regression models are not capable of handling data uncertainty. Indeed, in inverse problems of modeling and data mining, solutions can represent a remarkable sensitivity with respect to perturbations in the parameters which base on the data, and a computed solution can be highly infeasible, suboptimal, or both. Since, with increased volatility and further uncertainties, economical, environmental and financial crises translated a high "noise" within data into the related models, the events of recent years in the world have led to radically untrustworthy representations of the future, and robustification has started to attract more attention in many areas. Hence, we include the existence of uncertainty considering future scenarios into MARS and CMARS, and robustify them through *Robust Optimization* (*RO*) [14, 15], proposed to cope with data uncertainty. We

A. Özmen, *Robust Optimization of Spline Models and Complex Regulatory Networks*, Contributions to Management Science,
DOI 10.1007/978-3-319-30800-5_1

represent the new Robust (C)MARS (R(C)MARS) [96–98, 101] in theory and method, and apply R(C)MARS on some different kinds of datasets.

1.1 Purpose of the Study

The presence of noise and data uncertainty rises critical problems to be coped with on the theoretical and the computational side. In order to overcome that difficulty, new models have to be developed where optimization results are combined within real-life applications. For immunizing against parametric uncertainties, RO, developed by Ben-Tal and Nemirovski [9, 10, 14], and El-Ghaoui et al. [36, 37], has gained in great importance as a modeling framework from both a theoretical and a practical point of view. RO aims to find an optimal or near optimal solution that is feasible for each possible realization of the uncertain scenarios.

In order to make MARS and CMARS models more generalized and prepared to deal not only with fixed but also random type of input data, we introduce the new methods called RMARS and RCMARS by further enhancing the MARS and CMARS methods to handle data uncertainty [96, 105]. Because of the computational effort which our robustification of (C)MARS easily needs, we also describe our new concept of a weak robustification that is called as WR(C)MARS. In our book, we focus on the polyhedral type of uncertainty which brings us back to CQP naturally. By using robustification in (C)MARS, we aim to reduce the estimation variance. Furthermore, we analyze Generalized Partial Linear Models (GPLMs), and we introduce a newly developed Robust (Conic) Generalized Partial Linear Model (R(C)GPLM) [99, 100, 102] using the contribution of a continuous regression model R(C)MARS and a parametric/discrete regression model Logistic/Linear Regression. A R(C)GPLM leads to reduce the complexity of (C)MARS consisting in the number of variables used in R(C)MARS algorithm.

In this book, the robust optimization technique of solving and optimizing the models having nonlinearity and uncertainty by using R(C)MARS is also discussed with an implementation on two-model regulatory systems (Target-Environment (TE) systems) that appear in the financial sector and in banking, in environmental protection, system biology, medicine and so on. Since practitioners in these fields need to be aware that evaluation of probabilities based on history may be fundamentally inaccurate, uncertainty has importance for players in these sectors. The practice of using models of risks in a world of uncertainty is one of the reasons for the recent environmental and financial crisis [38, 39]. We have presented a regression model by using splines for the entries of regulatory network and achieved a relaxation by means of robust and continuous optimization, especially, Robust Conic Quadratic Programing (RCQP). That model of a TE regulatory system allows us to determine the unknown system parameters from uncertain measurement data by applying interior point methods [114, 115]. In case of uncertain data, polyhedral sets are used to encompass errors, what refers us to particular robust counterpart programs.

We are interested in the multicriteria **tradeoff** (antagonism) between ***accuracy*** and ***robustness***. In the line of our research [96–98, 101], robustness has become, in some sense, an extension of stability or regularity. Stability also means a *small complexity* of the model, or: a *small variance* of the estimation. Through R(C)MARS, we have included uncertainty into our regression and classification modeling not only in the output variable but especially in the input variables also. This type of uncertainty is typical for real-life applications. So, in RCMARS, there exists a double way of robustification: (i) The robustification is performed with respect to the input variables and output variable, all of them with their own uncertainty sets. (ii) The regularization in integral form that expresses itself by the involvement of the (squared) first- and second-order partial derivatives of the multivariate basis functions; after discretization of the integrals, we reach a TR problem with first- and second-order complexity terms. Then, this TR problem is turned into a CQP problem. In our RMARS, we only have the robustification step (i), whereas the fine-tuning option (ii) dropped. We underline the core importance of the target of numerical accuracy (precision) as a central goal in of our tradeoff that it establishes together with *regularity* and *robustness*, respectively. Within the RCMARS concept and its RCQP optimization program, through the "control parameter" which is represented by the upper bound of the complexity term, one can regulate and "tune" the importance that one assigns for the stability (lack of complexity) goal and, by this, for the antagonistic precision target. Moreover, we got the promise of gaining from the "power" of RO to our R(C)MARS [96, 101]. We demonstrate the well performance of our models with numerical experiences, simulation studies and real-world applications.

1.2 The Significance of Uncertainty

Since the global economic crisis has caused the necessity for an essential restructuring of the approach to risk and regulation, core elements of a new global regulatory-framework have become needed to establish in order to make the financial system more robust and suitable for serving the requirements of the real economy. For this reason, many scientists try to find ways to measure the probability of financial calamities, natural disasters and other catastrophes [39]. They draw attention to the difference between ***known risks*** and ***uncertainty***. The problem to be thought about is that most economists and other risk modelers do not separate uncertainty from risk. Economic models suppose that the financial world contains known risks that can be evaluated depending on prior behavior of stock markets and other elements of the monetary system. Nevertheless, there is genuine *uncertainty*, which is the impossibility of knowing exactly what the future keeps even in a probabilistic sense, as well as the *risk* that is the range of probabilities of outcomes pointed out by past events, which may serve as an unreliable guide for the future in an uncertain environment [22, 38, 48]. In other words, there are some sources of uncertainty: the data of the problem are not exactly known or may not be exactly measured, or the exact solution of the problem may not be implemented because of inherent inaccuracy of the devices [21], and data uncertainty results in uncertain constraints and in the objective function. This means that the known statistical models may not give trustworthy results.

Uncertainty is often presented in the sectors of energy, economics, finance, insurance, but also in high-tech and the environmental studies. It is one of the characteristic properties in these sectors since the given data, in both input and output variables, are affected with "noise" of various kinds, and the scenarios which represent the developments in time, are not deterministic either. Traditional approaches to optimization under uncertainty such as stochastic programming [64, 111], chance-constrained programming [26] or stochastic dynamic programming [17] generally have most serious numerical difficulties as the models in these areas are large and complex, already in their deterministic formulation. Therefore, as an alternative to traditional methods, RO is introduced to handle the complexity issue in adopting a non-probabilistic formulation of the uncertainty. RO does not have recourse to the calculus of probability that makes it immune against the curse of dimensionality and computational intractability [42].

In this book, the existence of uncertainty has a strong impact on the way of modeling which, then, becomes the basis of regression and classification and, eventually, of decision making. In this way, the uncertainty phenomenon enters all parts of the model and its mathematical treatment, and one of the prominent techniques to address this situation is RO. In fact, it refers to *worst-case scenarios*. In our study, we have at hand control parameters in order to regulate the amount of skepticism and optimism or, in other words, *risk-aversion* and *risk-friendliness* in the modeling process. Here, *risk* expresses itself in terms of variance, namely, the estimation variance, as we shall explain. We wish to underline that by all these considerations and measurements, with our robustification we are going much beyond of the concept of regularization which just relates to the output data and the complexity of the modeling functions.

1.3 Robust Optimization

Optimization has been a leading methodology in many fields such as engineering, finance and control design, and most applications suppose complete knowledge of the data which are underlying the optimization problem. In other words, it is assumed that to develop a model, the input data are known exactly (fixed). Nevertheless, solutions to optimization problems can represent a significant sensitivity to perturbations in the parameters of the problem. Optimization affected by parameter uncertainty is a focus of the mathematical programming community and a necessity to tackle uncertain data arises to develop models where optimization results are combined within real-world applications [18, 20].

RO is a method to address data uncertainty in optimization problems. The RO approach aims to make the optimization model robust, considering constraint violations by solving *robust counterparts* (*RCs*) of these problems in prespecified uncertainty sets for the uncertain parameters. These counterparts are solved for the *worst-case* realization of the uncertain parameters based on uncertainty sets for the random parameters [40].

RO has gained a lot of attention from both a theoretical and a practical point of view as a modeling framework for "immunizing" against parametric uncertainties in mathematical optimization. It is a methodology to model, evaluate and solve optimization problems in which the data are uncertain and only known to belong to some uncertainty set. RO purposes to receive an optimal or near optimal solution that is feasible for each possible realization of the uncertain scenarios [14, 146].

In this study, we work on R(C)MARS with the help of an RO approach which makes (C)MARS robust in view of constraint violations by solving RCs of these problems within uncertainty sets. Although these counterparts are solved for the worst-case realization of those uncertain parameters based on suitably defined uncertainty sets, we shall weaken the concept of "worst case" for reasons of computational complexity. Moreover, we have at hand control parameters which help to assess risk-friendliness- vs. -aversity. We study on robustification in terms of *polyhedral uncertainty* which enables us to return back to standart CQP naturally [96, 98, 101].

1.4 Complex Multi-Modal Regulatory Networks

The identification of the underlying network topology permits us to gain insights into the regulating effects and the hidden relationships between the variables. Many theoretical contributions from various disciplines concentrate on the analysis of such systems. Nevertheless, the identification of regulatory networks from real-world data is still a challenge in mathematics. This shows even more promise as the technical developments of the last decades have obtained a large number of data that are still waiting for a deeper analysis [27, 47, 56, 59, 61, 69, 109, 117, 132]. TE regulatory systems arise in many application areas in which they are more and more referred to as gene-environment or eco-finance networks. TE regulatory networks can be analyzed as gene-environment networks, for instance, to define the complex interactions between genes and other components of cells and tissues in modeling and prediction of gene-expression patterns from microarray experiments, regarding in a wider frame. The target variables are the expression values of the genes, whereas radiation, toxins, transcription factors, etc., additionally become environmental items [70, 71, 73, 78, 133, 135, 139, 143, 144]. Today, it is obviously understood that environmental factors comprise an essential group of regulating components and the performance of the models may be significantly improved by including these additional variables. The benefit of such a refinement has been shown, for example, in [141], where prediction and classification performances of supervised learning methods of the most complex-genome-wide human disease classification can be made better by taking into account environmental aspects.

TE regulatory networks may be extended with eco-finance networks (*eco* abbreviating 'ecology') by the important example in the area of CO_2-emissions-control; another example consists of operational planning and portfolio optimization for natural gas transportation systems. The interdisciplinary implications in economics, technology and Operational Research can be successfully explained by these kinds of network models [72, 78, 139, 144]. Furthermore, TE regulatory networks may be applied to an extension of the Technology-Emissions-Means (TEM) Model [110] that was prepared with the occasion of the Kyoto protocol [75]. The TEM model leads to a simulation of the cooperative economic behavior of countries or enterprises with the purpose decreasing the greenhouse gas emissions. Here, the target variables are the emissions which some countries have to diminish, and the financial expenditures act as additional environmental items [66, 109, 110]. There exist many other examples from biology and life sciences, which refer to TE-regulatory systems, with environmental effects being strongly included. Among them are, e.g., metabolic networks [23, 108, 140], immunological networks [56], social- and ecological networks [49]. Modeling and prediction of such regulatory systems and the problem of identifying the regulating effects and interactions between the targets and other components of the network have a significant importance in the mentioned areas.

The comparison of measurements and predictions of the TE regulatory networks lead to a regression models for parameter estimation. In most of the applications from these fields, it is assumed that the input data are not random but known (fixed) in developing models. Additionally, the data can undergo small changes by variations in the optimal experimental design. Therefore, all of these conditions cause uncertainty in the objective function and in possible constraints, and they introduce some kinds of weaknesses to the methods, because real-life data involve uncertainty in the form of noise. Here, since the regression models of target-environment networks can be affected by noise and errors, presented by *intervals*, the uncertain multivariate states are in total represented by *polyhedra*, and accordingly, our newly developed robust modeling techniques R(C)MARS, which can handle random inputs is used. This employs the concept of robustness through RO problems.

1.5 Scope of the Book

This book is comprised of seven main chapters and four appendices. Briefly summarizing, the contents are organized as follows:

Chapter 1: Introduction of the book. The objectives and outlines of the study is given in this chapter.
Chapter 2: The background information about Multi-Model Regulatory Networks, Optimization and Regression is provided.

Chapter 3: Theory and approaches of R(C)MARS and R(C)GPLM method under Polyhedral Uncertainty are demonstrated here.
Chapter 4: Spline Regression Models for Multi-Model Regulatory Networks are introduced in theory and methods. (C)MARS results based on different datasets for the simulation are represented.
Chapter 5: Robust Optimization in Spline Regression Models for Multi-Model Regulatory Networks is introduced in theory and methodology. R(C)MARS results with different uncertainty scenarios for the numerical example are studied here.
Chapter 6: Real-world applications from different sectors are presented in this chapter.
Chapter 7: A conclusion and an outlook to further studies are stated in the last chapter.

Chapter 2
Mathematical Methods Used

In this chapter, we introduce some preliminaries related with our studies.

2.1 Optimization

2.1.1 Robust Optimization

Robust optimization (RO) has gained a lot of attention both from a theoretical and practical point of view as a modeling framework for immunizing against parametric uncertainties in mathematical optimization. It is a modeling methodology to process optimization problems in which the data are uncertain and is only known to belong to some uncertainty set. Robust optimization purposes to receive an optimal or near optimal solution that is feasible for every possible realization of the uncertain data [14, 146].

In the early 1970s, Soyster [121] was one of the first researchers to investigate explicit approaches to RO. This short note focused on robust linear optimization in the case where the column vectors of the constraint matrix were constrained to belong to ellipsoidal uncertainty sets. He suggested a linear optimization model to create a solution that was feasible for all input data such that each uncertain input data point could take any value from an interval but, this approach tended to find solutions that were over-conservative. Even though Falk [41] followed this a few years later with more work on inexact linear programs, the optimization community was relatively quiet on the issue of robustness until the work of Ben-Tal and Nemirovski [9–11] and El Ghaoui et al. [36, 37] in the late 1990s.

The RO approach makes the optimization model robust regarding constraint violations by solving *robust counterparts* of these problems within prespecified uncertainty sets for the uncertain parameters. These counterparts are solved for

A. Özmen, *Robust Optimization of Spline Models and Complex Regulatory Networks*, Contributions to Management Science,
DOI 10.1007/978-3-319-30800-5_2

the worst-case realization of those uncertain parameters based on appropriately determined uncertainty sets for the random parameters [40]. The general uncertain optimization problem is defined as follows:

$$\begin{aligned} &\underset{\boldsymbol{x}\in\mathbb{R}^n}{\text{minimize}}\ \boldsymbol{c}^T\boldsymbol{x} \\ &\text{subject to } f_i(\boldsymbol{x},\boldsymbol{\varphi}) \le 0\ (i=1,2,\ldots,m), \end{aligned} \tag{2.1}$$

where $f_i(\boldsymbol{x},\boldsymbol{\varphi})$ are given constraint functions, $\boldsymbol{\varphi} \in \mathbb{R}^K$ is a vector of uncertain parameters and $c \in \mathbb{R}^n$ as well. Ben-Tal and Nemirovski [9–11] and, independently, El Ghaoui et al. [36, 37] have taken a significant step forward on developing theory for robust optimization. Indeed, the uncertain optimization problem is a family of problems—one for each realization of $\boldsymbol{\varphi}$. In the RO framework the information related to $\boldsymbol{c}$ and $\boldsymbol{\varphi}$ are modeled as geometric uncertainty sets $U_1 \subset \mathbb{R}^q$ and $U_2 \subset \mathbb{R}^K$. Therefore, the family of problems of Eq. (2.1) is rewritten by its *robust counterpart* in the following form:

$$\begin{aligned} &\underset{\boldsymbol{x}}{\text{minimize}}\ \max_{\boldsymbol{c}\in U_1} \boldsymbol{c}^T\boldsymbol{x} \\ &\text{subject to } f_i(\boldsymbol{x},\boldsymbol{\varphi}) \le 0,\ \forall\ \boldsymbol{\varphi} \in U_2\ (i=1,2,\ldots,m), \end{aligned} \tag{2.2}$$

where U_1 and U_2 are given uncertainty sets. Let any minimal value be called z^*. The motivation is to find a solution of the stated problem in Eq. (2.2) that 'immunizes' the problem Eq. (2.1) against parameter uncertainty. Here, the objective function is guaranteed to be no worse than z^* and a solution of Eq. (2.2) is feasible to Eq. (2.1) for any realization of $\boldsymbol{\varphi} \in U_2$. Anyone of the two uncertainty sets U typically is a polytope or an ellipsoid or an intersection of such sets. In the robust optimization literature, a general form of uncertainty set, U, e.g., U, is given as follows:

$$U = \left\{ \boldsymbol{\varphi} = \overline{\boldsymbol{\varphi}} + \sum_{\kappa=1}^{q} \rho_\kappa \boldsymbol{\varphi}^\kappa \in \mathbb{R}^K \middle| \ \boldsymbol{\rho} \in Z \right\}, \tag{2.3}$$

where $\overline{\boldsymbol{\varphi}}$ is the nominal value of the uncertain vector $\boldsymbol{\varphi}$, the vectors $\boldsymbol{\varphi}^K$ are possible scenarios of it, and $\boldsymbol{\rho} = (\rho_1,\rho_2,\ldots,\rho_q)^T$ is a perturbation vector. The set Z determines what type of uncertainty set we have. These sets may be

$$\begin{aligned} \textit{box uncertainty set}: Z &= \left\{\boldsymbol{\rho} \in \mathbb{R}^q \middle|\ \boldsymbol{\rho} \ge 0,\ \boldsymbol{e}^T\boldsymbol{\rho} \le 1\right\}, \\ \textit{convex combination of scenarios}: Z &= \left\{\boldsymbol{\rho} \in \mathbb{R}^q \middle|\ -1 \le \rho_i \le 1\ (i=1,2,\ldots,q)\right\}, \\ \textit{ellipsoid uncertainty set}: Z &= \left\{\boldsymbol{\rho} \in \mathbb{R}^q \middle|\ \boldsymbol{\rho}^T\boldsymbol{\rho} \le 1\right\}. \end{aligned} \tag{2.4}$$

where $\boldsymbol{e} = (1,1,\ldots,1)^T \in \mathbb{R}^q$.

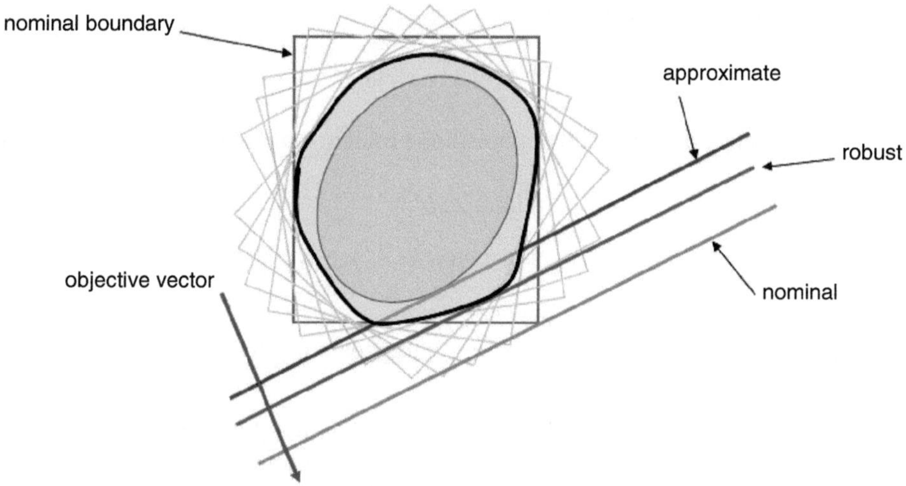

Fig. 2.1 Approximating a robust solution [35]

These sets yield useful models of uncertainty, which lead to tractable optimization problems [15]. For a visualization see Fig. 2.1.

2.1.2 *Conic Optimization*

A general *primal conic* (*CP*) optimization problem is a problem in the conic form

$$\begin{aligned} &\text{minimize } \boldsymbol{c}^T\boldsymbol{x} \\ &\text{subject to } \boldsymbol{A}\boldsymbol{x} = \boldsymbol{b},\ \boldsymbol{x} \in K, \end{aligned} \tag{2.5}$$

where K is a closed, pointed (which means that K contains no line), non-empty, convex cone, $\boldsymbol{x}$ is the design vector and $\boldsymbol{c}$ is a given vector of coefficients of the objective function. In fact we assume that K is some product of the following cones [12]:

(*i*) The *nonnegative orthant* $\mathbb{R}^n_+$. The non-negative orthant consists of all vectors in $\mathbb{R}^n_+$ whose elements are all non-negative: $\mathbb{R}^n_+ = \{\boldsymbol{x} | x_k \geq 0\ \forall k = 1, 2, \ldots, n\}$.
(*ii*) The *Lorentz* (or second order, or ice-cream) cone:

$$L^n := \left\{\boldsymbol{x} = (x_1, x_2, \ldots, x_n)^T \in \mathbb{R}^n \middle|\ x_n \geq \sqrt{x_1^2 + x_2^2 + \ldots + x_{n-1}^2}\right\}\ (n \in \mathbb{N} \setminus \{1\}).$$

(*iii*) The positive semidefinite cone:

$$L_+^n := \{A \in \mathbb{R}^{n\times n} \,|\, A = A^T,\, x^T A x \geq 0 \;\forall x \in \mathbb{R}^n\}. \tag{2.6}$$

A general *dual conic* (CD) optimization problem is a problem in the conic form

$$\begin{aligned} &\text{minimize } b^T y \\ &\text{subject to } c - A^T y \in K. \end{aligned} \tag{2.7}$$

Here, (CD) is just called the *dual problem* of (CP), the constraint is called a linear matrix inequality. For a conic problem, there exist the following properties of the duality [12]:

1. The value of the dual objective at every dual feasible solution is the value of the primal objective at every primal feasible solution (*weak duality*).
2. The following two properties are equivalent to each other:

   (*a*) The primal problem is strictly feasible and below bounded,
   (*b*) The dual problem is solvable.

3. The following two properties are equivalent to each other:

   (*c*) The dual problem is strictly feasible and bounded from below,
   (*d*) The primal problem is solvable.

   Strong Duality property: Whenever $(a)(\Leftrightarrow\ (b))$ or $(c)(\Leftrightarrow\ (d))$ is the case, the optimal values in the primal and the dual problems are equal to each other (*strong duality*):

$$Opt(CP) = Opt(CD).$$

4. The duality is *symmetric*: the problem dual to the dual is equivalent to the primal. There are different conic optimization problems considered and coped with such as Linear Programming (LP), Second-Order Cone Programming (SOCP), Semidefinite Programming (SDP). In all these cases, these conic optimization problems can be solved efficiently by an *interior Point Method* (*IPM*) [114, 115]. For our study, we will mainly focus on SOCP, also called as *Conic Quadratic Programming* (*CQP*) [12, 19]. Here, to find a solution for conic optimization problem, MOSEK [87], SeDuMi [125], SDPT3 [131] can be used as a solver.

2.1.2.1 Conic Quadratic Programming

CQP is the problem of minimizing a linear objective function subject to the intersection of an affine set and the direct product of quadratic cones of the form

$$L^n = \left\{ \boldsymbol{x} \in \mathbb{R}^n \middle| \, x_n^2 \geq \sum_{j=1}^{n-1} x_j^2, \quad x_{n-1} \geq 0 \right\}.$$

We recall that the quadratic cone is also known as the second-order (Lorentz or ice-cream) cone. Many optimization problems can be constructed in this form. Some examples are linear, convex quadratic and convex quadratically constrained quadratic optimization. Various applications of conic quadratic optimization are presented in [12, 79]. A conic optimization problem can be represented in the following form:

$$\begin{aligned} &\text{minimize } \boldsymbol{c}^T \boldsymbol{x} \\ &\text{subject to } \boldsymbol{A}\boldsymbol{x} = \boldsymbol{b}, \ \boldsymbol{x} \in K, \end{aligned}$$

associated with a cone K, represented as $K = L^{n_1} \times L^{n_2} \ldots \times L^{n_r} \subseteq E$. Canonically turning to inequalities rather than equalities, in general, a CQP is an optimization problem with linear objective function and finitely many 'ice-cream constraints' $\boldsymbol{b}_i - \boldsymbol{A}_i\boldsymbol{x} \underset{L^{n_i}}{\geq} \boldsymbol{0}$, defined by $\boldsymbol{b}_i - \boldsymbol{A}_i\boldsymbol{x} \in L^n$ $(i = 1, 2, \ldots, r)$. Therefore, a CQP problem can be written as [12]

$$\begin{aligned} &\underset{\boldsymbol{x}}{\text{minimize}} \ \boldsymbol{c}^T \boldsymbol{x} \\ &\text{subject to } \ \boldsymbol{b}_i - \boldsymbol{A}_i\boldsymbol{x} \underset{L^{n_i}}{\geq} \boldsymbol{0} \ \ (i = 1, 2, \ldots, r). \end{aligned}$$

If we subdivide the data matrix, $[\boldsymbol{A}_i, \boldsymbol{b}_i]$, as follows:

$$[\boldsymbol{A}_i, \boldsymbol{b}_i] = \begin{bmatrix} \boldsymbol{D}_i & \boldsymbol{d}_i \\ \boldsymbol{p}_i^T & q_i \end{bmatrix},$$

where $\boldsymbol{D}_i$ is of the size $((m_i - 1) \times \dim \boldsymbol{x})$, the problem can be written as follows:

$$\begin{aligned} &\underset{\boldsymbol{x}}{\text{minimize}} \ \boldsymbol{c}^T \boldsymbol{x} \\ &\text{subject to } \ \|\boldsymbol{D}_i\boldsymbol{x} - \boldsymbol{d}_i\|_2 \leq \boldsymbol{p}_i^T \boldsymbol{x} - q_i \ \ (i = 1, 2, \ldots, r). \end{aligned}$$

That is the most explicit form which is used. In that form, $\boldsymbol{D}_i$ are matrices with the row dimensions being the dimension of $\boldsymbol{x}$, $\boldsymbol{d}_i$ are vectors of the same dimensions as the column dimensions of the matrices $\boldsymbol{D}_i, \boldsymbol{p}_i$ are vectors of the same dimensions as $\boldsymbol{x}$, and q_i are real numbers.

2.1.2.2 Interior Point Methods

Convex optimization problems like semidefinite programming, geometric programming and, in particular, CQP problems are very essential in data mining and classical polynomial-time algorithms may be employed to solve these kinds of problems. Nevertheless, these algorithms have some disadvantages since they use local information on the objective function and the constraints. Therefore, *Interior Point Methods* (*IPMs*) [92], firstly introduced by Karmarkar [62], are employed to solve 'well-structured' convex problems, like CQP problems. There has been done comprehensive research on interior-point methods for linear optimization. One result of this research is the development of a primal-dual interior-point algorithm [65, 88] that is highly efficient both in theory and in practice [3, 81]. Consequently, some authors have studied to drive this algorithm for other problems. An important work in this direction is the paper of Nesterov and Todd [93] which represents that the primal-dual algorithm keeps its theoretical efficiency when the nonnegativity constraints are replaced by a convex cone as long as the cone is homogeneous and self-dual, or in the terminology of Nesterov and Todd, a self-scaled cone [4]. It has subsequently been implied by Güler [52] that the only interesting cones having this property are direct products of the quadratic cone and the cone of positive semi-definite matrices. For our study, we mainly focus on conic quadratic optimization and on an algorithm for this class of problems.

For CQP, many authors have already worked algorithms. In particular, Monteiro and Tuschiya [89, 129] have analyzed the complexity of different variants of the primal-dual algorithm. Schmieta and Alizadeh [118] have represented that many of the polynomial algorithms developed for semidefinite optimization [124] may immediately be translated to polynomial algorithms for conic quadratic optimization [4]. Sturm [125] has reported that his code SeDuMi may solve conic quadratic and semidefinite optimization problems. We take into consideration an optimization problem given by [4, 12]

$$\underset{x \in \wp}{\text{minimize}}\, \boldsymbol{c}^T \boldsymbol{x},$$

where $\wp \subseteq \mathbb{R}^n$. Here, IPMs base on the interior point of the feasible set $\wp$. We suppose that this feasible set is closed and convex. An interior penalty function (*barrier*) $F(\boldsymbol{x})$ is selected, well defined, smooth and strongly convex, in the interior of $\wp$ and blowing up as a sequence from the interior int $\wp$ approaches a boundary point of $\wp$:

$$\boldsymbol{x}_r \in \operatorname{int} \wp \ (n \in \mathbb{N}_0),\ \lim_{r \to \infty} \boldsymbol{x}_r \in \partial \wp \ \Rightarrow\ F(\boldsymbol{x}_r) \to \infty \ (r \to \infty).$$

Now, we take into account one parametric family of functions generated by our objective and interior *penalty function*

$$F_p(\boldsymbol{x}) := p\boldsymbol{c}^T + F(\boldsymbol{x})|\text{int } \wp \to \mathbb{R}.$$

The penalty parameter p is supposed to be nonnegative. Under mild regularity assumptions [4],

(*i*) every function $F_p(\cdot)$ attains its minimum over the interior of $\wp$, the minimizers $x_*(p)$ being unique;
(*ii*) the central path $x_*(\cdot)$ is a smooth curve, and all of the variables p, its limiting points (as $p \to \infty$), belong to the set of optimal solution of above optimization problem.

These algorithms have the advantage of employing the structure of the problem, of allowing better complexity bounds and exhibiting a much better practical performance. In the so-called *primal-dual IPMs*, both the primal and the dual problems and their variables are regarded, the joint optimality conditions perturbed, parametrically solved and followed towards a solution along a *central path*.

2.1.3 Robust Conic Optimization

For all (or most) possible realizations of the data, the solution should satisfy the real constraints despite of the data uncertainty. Such a solution is called a *robust feasible solution*. The problem of receiving an optimal robust solution is called the robust counterpart of the original problem. Indeed, it is the problem of minimizing the objective function over the set of robust feasible solutions. In this study, we deal with an uncertain conic problem which has the following form [19]:

$$\begin{aligned} &\underset{\boldsymbol{x}}{\text{minimize}}\ \boldsymbol{c}^T\boldsymbol{x}, \\ &\text{subject to}\quad \boldsymbol{A}_k\boldsymbol{x} - \boldsymbol{b}_k \in K_k \quad (k = 1, 2, \ldots, N), \end{aligned}$$

where K_k $(k = 1, 2, \ldots, N)$ are closed, pointed, non-empty, convex cones, and $\boldsymbol{A}, \boldsymbol{b}, \boldsymbol{c}$ are subject to data uncertainty. It is necessary that the robust counterpart is computationally tractable, that is, solvable in polynomial time with respect to the problem size for RO, which is an applicable methodology for real-life large-scale problems [36, 120]. We note that tractability of the robust counterpart depends on the original optimization problem and the uncertainty set considered.

The robust optimization problem can be solved efficiently when the uncertainty set has a special shape. These special shapes for uncertainty sets can be either ellipsoidal or polyhedral. If ellipsoidal uncertainty sets are applied, the robustification is more successful than the employing of polyhedral uncertainty sets. However, the complexity of optimization problems increases when an ellipsoidal, rather than a polyhedral uncertainty set is applied. Indeed, our *robust CQP* (*RCQP*) problem becomes a problem of *Semidefinite Programming* (*SDP*) [8, 13, 37] under ellipsoidal uncertainty.

To not increase the complexity of optimization problems involved, in this study, we only focus on polyhedral uncertainty with different uncertain scenarios. We study our RCQP problem (robust second-order optimization problem (RSCOP)) and we shall find out that it remains CQP. Consequently, we will guarantee polyhedral uncertainty sets by an interval concept for input and output data in our model; our RCQP problem will be traced back directly as CQP programs.

2.1.4 Multi-Objective Optimization

In general optimization problems, there is a single objective function and the aim is to find a solution which optimizes the objective function value subject to some constraints by using single-objective optimization method. Nevertheless, most real-world problems have several objectives, and decisions must be made by regarding these objective functions at the same time [123]. When an optimization problem includes more than one objective function, this problem is called as *Multi-Objective Optimization* (*MOO*) problem that has the task of finding one or more optimum solutions [32]. If optimization problems contain multiple objectives, we cannot use single-objective optimization methods. In fact, different objectives are commonly conflicting with each other. Therefore, a solution which performs well in one objective cannot do as good as in the other objectives [32]. There exist several solutions that do not perform suitably in all objectives. It is not clear which of these solutions are better until the decision maker computes them. An MOO problem can be written as following form [123]:

$$\underset{\boldsymbol{x}}{\text{minimize}}\ (f_1(\boldsymbol{x}), f_2(\boldsymbol{x}), \ldots, f_q(\boldsymbol{x}))^T \text{such that}\ \boldsymbol{x} \in X,$$

where $\boldsymbol{x} \in \mathbb{R}^n$ is a feasible solution and $X \subseteq \mathbb{R}^n$ is the set of all feasible solutions. In this problem, there are q objective functions to be minimized. Sometimes the MOO problem is *symbolically* written with a $(q \times n)$-matrix $\boldsymbol{A}$, where the ith row of $\boldsymbol{A}$ corresponds to the ith objective function, $f_i(\boldsymbol{x})$.

The point $\boldsymbol{y} = (y_1, y_2, \ldots, y_q)^T \in \mathbb{R}^q$ such that $\boldsymbol{y} = \boldsymbol{A}\boldsymbol{x}$ is the outcome of the solution $\boldsymbol{x} \in X$. The set X is called decision space, and $Y = \{\boldsymbol{y} \in \mathbb{R}^q \,|\, \boldsymbol{y} = \boldsymbol{A}\boldsymbol{x}, \boldsymbol{x} \in X\}$ is called the objective (criterion) space. A point $\boldsymbol{x}$ is called to *dominate point* $\boldsymbol{x}'$ if and only if the corresponding $y_q \leq y'_q$ for all q and $y_q < y'_q$ for at least one q. If there is no $\boldsymbol{x}' \in X$ such that $\boldsymbol{x}'$ dominates $\boldsymbol{x}$, then $\boldsymbol{x}$ is called *non-dominated* or *efficient*. The complete set of non-dominated solutions is also known as the *Pareto-optimal* set.

2.1.5 *Optimization Softwares*

It is important to make distinction between optimization solvers (optimizer) and optimization modeling languages [40]. An optimization solver is a software that carries out numerical routines to obtain the optimal solution of an optimization problem. Optimization modeling languages appeared as user-friendly platforms that permit the user to specify optimization problems. AMPL and GAMS, communicating with a diversified amount of solvers, are two popular modeling languages. Also, there are lots of languages which give modeling interfaces for particular types of optimization problems or solvers [40]. For example, YALMIP [80] let Matlab users to preprocess SDPs and CQPs. Then, these are passed to semidefinite solvers like SDPT3 and SeDuMi.

SDPT3 [131] and SeDuMi [125] can handle linear constraints, quasi-convex quadratic constraints and positive semidefinite constraints. Two of them use a primal-dual interior points method implied as the centering-predictors-correctors method, and may exploit sparse matrix structure, making them very efficient [124]. For these semidefinite programming solvers, creating the inputs may be very time consuming, and can need substantial background in optimization modeling. YALMIP and PROF which are obtained as layers on top of these solvers in Matlab permit for intuitive formulation of SDPs and SOCPs, and help the user retrieve the results from the solvers very easily [40].

MOSEK is a useful optimizer for linear, quadratic and convex quadratically constrained optimization problems, well-known for speed and numerical stability [40]. It enables solvers for the optimization problems which have the types of the linear, conic quadratic (CQ), convex quadratic, general convex and mixed integer. MOSEK optimization tool consists of interfaces to make it easy to employ the functionality of MOSEK from programming languages such as C, C++, MATLAB Toolbox, Java, NET, and Python [87]. MOSEK technique has some technical benefits and an optimization tool to solve large-scale mathematical optimization problems, but the problem size is only limited by the available memory. MOSEK is of an interior-point optimizer with basis identification and it is well known owing to its excellent speed and stability [87]. The software uses problem sparsity and structure automatically to receive the best possible efficiency. It also has both primal and dual simplex optimizers for Linear Programming (LP) and corrects sensitivity analysis for linear problems. It has an efficient presolver to decline problem size before optimization. It can tackle primal and dual infeasible problems in a systematic way [87]. Furthermore, MOSEK contains tools for infeasibility diagnosis and repair and, it may read and write industry standard formats such as MPS, LP and XML.

2.2 Dynamical System of Complex Multi-Modal Regulatory Networks

Dynamic systems abound in the real-life practical environment as biological, mechanical, electrical, civil, aerospace, medicine, environmental sciences, finance and economy and a variety of other systems. Understanding the dynamic behavior of these systems is of primary interest to scientists as well as engineers. The availability of large data sets now allows to gain deeper insights in the dynamic behavior of complex systems and opens promising avenues for further scientific progress. These systems often involve two different kinds of data sets in form of certain key or target variables and additional environmental variables. For a deeper analysis one has to describe and investigate the interactions and regulating effects between data items of interest and the environmental items, encoded in the regulation-network. Modeling and anticipation of such systems and the problem of identifying regulating effects and interactions between the targets and the other components of the network have a remarkable significance in the mentioned areas [14, 68]. As these models are based on real-world data, errors and uncertainty have to be considered.

Examples

(*a*) The models under consideration is developed in the context of modeling and prediction of gene-expression patterns [133, 135, 138, 139, 142, 145]. In these gene-environment networks, the target variables represent the expression levels of the n genes, whereas the m environmental factors denote external items (e.g., radiation or toxins).

(*b*) TE regulatory-networks may be extended with eco-finance networks ('eco' abbreviating 'ecology') with an important example in the area of CO_2-emissions-control; another example of operational planning and portfolio optimization for natural gas transportation systems. In [66, 109, 110, 139], the Technology-Emissions-Means Model (in short: TEM-model) is investigated, which lets a simulation of the cooperative economic behaviour of countries/enterprises with the purpose of a reduction of CO_2-emissions. Here, the target variables are the emissions that the actors wish to decrease and the required financial means act as additional environmental items.

2.2.1 *Time-Continuous Regulatory Networks*

With regard to different stages of modeling we can categorized two situations:

(*i*) Networks with n targets (by disregarding the environmental factors),
(*ii*) Networks with n targets as well as m environmental factors.

For this, we divide the vector $\mathbb{E}$ of concentration levels into two parts and construct $\mathbb{E} = (\mathbb{E}_1, \mathbb{E}_2, \ldots, \mathbb{E}_n, \mathbb{E}_{n+1}, \mathbb{E}_{n+2}, \ldots, \mathbb{E}_{n+m})^T$, where $\mathbb{E}_1, \mathbb{E}_2, \ldots, \mathbb{E}_n$

refer to the n targets and $\mathbb{E}_{n+1}, \mathbb{E}_{n+2}, \ldots, \mathbb{E}_{n+m}$ to the m environmental factors, respectively. If we deal with models of type (i), E_i indicates the expression level of target i and E denotes the first n coordinates of the $d = n + m$-vector $\mathbb{E}$ [145].

A dynamical system of n targets (without any environmental factors) can be stated by the continuous differential equation

$$\dot{\boldsymbol{E}} = \boldsymbol{A}(\boldsymbol{E})\boldsymbol{E}, \tag{2.8}$$

where the matrix $\boldsymbol{A}$ can depend on $\boldsymbol{E}$ (cf. [133, 138]).

To involve environmental factors into continuous model under the presence of noise and uncertainty we extended in [133, 138] the model from [47] and provided the continuous equation, equipped with an initial vector:

$$\dot{\mathbb{E}}^{(k+1)} = \mathbb{A}^{(k)}\mathbb{E}^{(k)}, \;\; \mathbb{E}(t_0) = \mathbb{E}^{(0)}. \tag{2.9}$$

The associated system matrix $\mathbb{A}(\mathbb{X})$ is a $(d \times d)$-matrix described by a family of functions which have unknown parameters. Now, intervals present uncertainty in the states, partially caused by uncertainty in the interactions. We refer to the interactions between the targets, to the effects between the environment and the targets, or between environmental factors. The initial value $\mathbb{E}^{(0)} = (\mathbb{E}_1^{(0)}, \mathbb{E}_2^{(0)}, \ldots, \mathbb{E}_d^{(0)})^T$ contains the interval-valued levels obtained by the first measurement, $\bar{E}(t_0) = \bar{E}^{(0)}$. Since this can result in a large and highly interconnected network, we need to restrict on an approximate model and network. Here, polynomial, trigonometric, exponential but otherwise logarithmic, hyperbolic, spline, etc., entries present any kind of a prior information, observation or assumption, in terms of growth, cyclicity, piecewise behavior, etc. In this book, we analyze regulatory systems with *spline* entries as an advanced case.

2.2.2 *Time-Discrete Regulatory Networks*

The time-discrete TE regulatory systems under consideration consist of n targets and m environmental factors. The expression values of the target variables are given by the vector $\boldsymbol{X} = (X_1, X_2, \ldots, X_n)^T$ and the vector $\boldsymbol{E} = (E_1, E_2, \ldots, E_m)^T$ denotes the states of the environmental variables. The intricate interactions and synergistic connections between variables—targets as well environmental factors—of the regulatory system depend on four types of regulating effects, respectively [70, 73]:

(TT) target variable $\rightarrow$ target variable,
(ET) environmental factor $\rightarrow$ target variable,
(TE) target variable $\rightarrow$ environmental factor, and
(EE) environmental factor $\rightarrow$ environment variable.

Predictions of the time-dependent states of targets $\boldsymbol{X}_j$ and environmental factors $\boldsymbol{E}_i$ can be calculated through the following parametrized time-discrete model:

$$\begin{aligned} X_j^{(k+1)} &= \alpha_{j0}^{\mathrm{T}} + \left(\boldsymbol{X}^{(k)}\right)^T \alpha_j^{\mathrm{TT}} + \left(\boldsymbol{E}^{(k)}\right)^T \alpha_j^{\mathrm{ET}}, \\ E_i^{(k+1)} &= \alpha_{i0}^{\mathrm{E}} + \left(\boldsymbol{X}^{(k)}\right)^T \alpha_i^{\mathrm{TE}} + \left(\boldsymbol{E}^{(k)}\right)^T \alpha_i^{\mathrm{EE}}, \end{aligned} \tag{2.10}$$

with $k \in \mathbb{N}_0$. Here, $\alpha_j^{\mathrm{TT}}, \alpha_j^{\mathrm{ET}} \in \mathbb{R}^n$ and $\alpha_i^{\mathrm{TE}}, \alpha_i^{\mathrm{EE}} \in \mathbb{R}^m$ stand for the vectors of parameters and $\alpha_{j0}^{\mathrm{T}}, \alpha_{i0}^{\mathrm{E}} \in \mathbb{R}$ are intercepts, respectively. The initial vectors $\boldsymbol{X}^{(0)}$ and $\boldsymbol{E}^{(0)}$ can be given by the first measurements of targets and environmental factors: $\boldsymbol{X}^{(0)} := \tilde{\boldsymbol{X}}^{(0)}$ and $\boldsymbol{E}^{(0)} := \tilde{\boldsymbol{E}}^{(0)}$.

The *regulatory model* (*RM*) in Eq. (2.10) depends on $(n+m)(n+m+1)$ unknown parameters. These parameters have to be assessed on the basis of noisy measurements. The fundamental idea of our regression analysis is to compare the predictions of (RM) with the (uncertain) states of targets $\tilde{\boldsymbol{X}}^{(k)} = (\tilde{X}_1^{(k)}, \tilde{X}_2^{(k)}, \ldots, \tilde{X}_n^{(k)})^T \in \mathbb{R}^n$ and environmental observations $\tilde{\boldsymbol{E}}^{(k)} = (\tilde{E}_1^{(k)}, \tilde{E}_2^{(k)}, \ldots, \tilde{E}_m^{(k)})^T \in \mathbb{R}^m$ $(k = 0, 1, \ldots, N)$ obtained from measurements at sampling times $t_0 < t_1 < \ldots < t_N$. By inserting these measurements into model (RM) we obtain the following predictions:

$$\begin{aligned} \widehat{X}_j^{(k+1)} &= \alpha_{j0}^{\mathrm{T}} + \left(\tilde{\boldsymbol{X}}^{(k)}\right)^T \alpha_j^{\mathrm{TT}} + \left(\tilde{\boldsymbol{E}}^{(k)}\right)^T \alpha_j^{\mathrm{ET}}, \\ \widehat{E}_i^{(k+1)} &= \alpha_{i0}^{\mathrm{E}} + \left(\tilde{\boldsymbol{X}}^{(k)}\right)^T \alpha_i^{\mathrm{TE}} + \left(\tilde{\boldsymbol{E}}^{(k)}\right)^T \alpha_i^{\mathrm{EE}}, \end{aligned} \tag{2.11}$$

where $k = 0, 1, \ldots, N-1$. We refer to initial values $\boldsymbol{X}_j^{(0)} := \tilde{\boldsymbol{X}}_j^{(0)}$ and $\boldsymbol{E}_i^{(0)} := \tilde{\boldsymbol{E}}_i^{(0)}$, as we define the vectors $\tilde{\boldsymbol{X}}^{(k)} = \left(\tilde{X}_1^{(k)}, \tilde{X}_2^{(k)}, \ldots, \tilde{X}_n^{(k)}\right)^T$ and $\tilde{\boldsymbol{E}}^{(k)} = \left(\tilde{E}_1^{(k)}, \tilde{E}_2^{(k)}, \ldots, \tilde{E}_m^{(k)}\right)^T$, where $k = 0, 1, \ldots, N; i = 1, 2, \ldots, m; j = 1, 2, \ldots, n$.

If now the entries of the matrices encoding regulatory network are specified by spline functions for being more flexible in approximating the data, and if we encounter interaction between the input variables, then this leads us to employ models that will be based on (R)MARS and (R)CMARS. Here, splines, as function of the input variable, are piecewise polynomials. If we only used polynomials, then they would generally converge to plus or minus infinity while the absolute values of the input variables grow large.

Since real-world processes usually stay in bounded margins even though these bounds are very large, polynomials would require being of a high degree to turn around or oscillate enough to stay in that margin. However, it is not easy to work with high-degree polynomials as the real-world problems are multivariate and this may imply multiplication effects. Instead of this, using splines lets us keep the degree of the polynomial pieces very low in each dimension. Indeed, splines are quite flexible, such to say, elastic. We frequently call them smoothing splines even, since they smoothly approximate the discrete data. Therefore, in this book, we analyze time-discrete TE regulatory systems with spline entries and introduce new regression and classification models that allow us to define the unknown system parameters by applying the (R)MARS and (R)CMARS techniques.

2.3 Inverse Problems and Parameter Estimation

An *Inverse Problem* involve to use the actual result of some measurements to figure out the values of the parameters that characterize the system. In an inverse problem, one has necessity to make explicit any available a priori information on the model parameters. One also needs to be careful in the representation of the data uncertainties [3, 126].

Parameter estimation is one of the main tasks of the scientist or engineer. Mathematical modeling via parameter estimation is one of the approaches that provides a deeper understanding of the characteristics of a regarded system. These parameters usually defined the stability and control behavior of the system. Therefore, prediction of these parameters from input-output data of the system is an essential step in the analysis of the dynamic system. Indeed, analysis refers to the process of constructing the system response to a specific input, given the knowledge of the model representing the system. Hence, in this process, knowledge of the mathematical model and its parameters is of primer significance. Our problem of parameter estimation belongs to the class of "inverse problems" in which the knowledge of the dynamical system is derived from the input data and the associated derivative of the system [16].

Most attention is drawn to the detailed definition of methods for parameter estimation, involving ordinary and weighted *least-squares* (*LS*) and maximum likelihood with and without prior information. Least-squares estimation (LSE) is widely preferred to use for solving inverse problems because they enable to the easiest computations [16]. The only drawback of these methods is their lack of robustness, i.e., their strong sensitivity to a small number of large errors (outliers) in a data set. To employ the LS method, the model should be written on the regression model of the next chapter in Eq. (2.12).

2.3.1 Least-Squares Estimation

In this section, we consider multiple linear regression model to apply LS method. We start with describing the multiple regression model and then, we give the LS method to estimate the parameters of the multiple linear regression model.

In general, the response variable Y may be related to p regressor variables. With the observations presented by the data $(\boldsymbol{x}_k, y_k)$ $(k = 1, 2, \ldots, N)$, the form of the models is follows:

$$Y_k = \alpha_0 + \alpha_1 x_{k,1} + \alpha_{k,2} x_2 + \ldots + \alpha_p x_{k,p} + \varepsilon_k, \tag{2.12}$$

are called a *multiple linear regression models* with p regression variables. The parameter α_0 means the intercept and the other parameters α_j $(j = 1,2,\ldots,p)$ are the regression coefficients. To select the **best-fitting** line for a set of data, the unknown parameters of the multiple linear regression model, α_0, α_1, …, α_p should be estimated.

LS method is widely applied to predict the parameters in regression models and describe the statistical properties of estimates. Assume that $N \geq p$ observations on the response variable are given as $y_1, y_2, \ldots, y_N$. For each observed response y_k, we have an observation on each dependent variable and let $x_{k,j}$ indicate the pth observation of variable $\boldsymbol{x}_j$ $(j = 1,2,\ldots,p;\ k = 1,2,\ldots,N)$. Here, we firstly suppose that true relationship between the dependent variable and independent variables are linear. We also suppose that the noise term ε_k in the model has $E(\varepsilon_k) = 0$ and $V(\varepsilon_k) = \sigma^2$ and that the ε_k are uncorrelated random variables [85, 86]. We may write the model of Eq. (2.12) based on observations included as

$$\begin{aligned} y_k &= \alpha_0 + \alpha_1 x_{k,1} + \alpha_2 x_{k,2} + \ldots + \alpha_p x_{k,p} + r_k, \\ &= \alpha_0 + \sum_{j=1}^{p} \alpha_j x_{k,j} + r_k \quad (k = 1,2,\ldots,N). \end{aligned} \tag{2.13}$$

The estimation method of least-squares selects the coefficients α_j in Eq. (2.13) provided that the sum of the squares of the errors, called *residuals*, r_k is minimized. The least-squares function is

$$L = \sum_{k=1}^{N} r_k^2 = \sum_{k=1}^{N} (y_k - \alpha_0 - \sum_{j=1}^{p} \alpha_j x_{k,j})^2. \tag{2.14}$$

Turning this into the matrix notation, the least-squares estimators should satisfy

$$\frac{\partial L}{\partial \boldsymbol{\alpha}}(\hat{\boldsymbol{\alpha}}) = -2\boldsymbol{X}^T \boldsymbol{y} + 2\boldsymbol{X}^T \boldsymbol{X} \hat{\boldsymbol{\alpha}} = \boldsymbol{0}, \tag{2.15}$$

which simplifies to

$$\boldsymbol{X}^T \boldsymbol{X} \hat{\boldsymbol{\alpha}} = \boldsymbol{X}^T \boldsymbol{y}. \tag{2.16}$$

Equation (2.16) is the matrix form of the least-squares normal equations. To solve the normal equations, multiply both sides of Eq. (2.16) by the inverse of $\boldsymbol{X}^T\boldsymbol{X}$, which exists if $N \geq (p+1)$ and the design matrix $\boldsymbol{X}$ has full rank. In this form it is obvious that $\boldsymbol{X}^T\boldsymbol{X}$ is a symmetric $((p+1)\times(p+1))$-matrix and $\boldsymbol{X}^T\boldsymbol{y}$ is a column $((p+1)\times 1)$-vector. The diagonal elements of $\boldsymbol{X}^T\boldsymbol{X}$ are the sums of squares of the elements in the columns of $\boldsymbol{X}$, and the off-diagonal elements are the sums of cross products of the elements in the columns of $\boldsymbol{X}$ and the observations y_k [86]. The fitted regression model is

$$\widehat{\boldsymbol{Y}} = \boldsymbol{X}\hat{\boldsymbol{\alpha}}. \tag{2.17}$$

In scalar notation, the fitted model is

$$\hat{y}_k = \hat{\alpha}_0 + \sum_{j=1}^{p} \hat{\alpha}_j x_{k,j} \ (k = 1, 2, \ldots, N).$$

The difference between the real observation y_i and the corresponding predicted value $\hat{y}_i$ is the residual (estimation errors), $r_k = y_k - \hat{y}_k$. The $(N \times 1)$-vector of residuals is implied by

$$\boldsymbol{r} = \boldsymbol{y} - \hat{\boldsymbol{y}}. \tag{2.18}$$

To develop an estimator of the parameter, σ^2, take into account the sum of squares of the residuals [86]:

$$SS_E = \sum_{k=1}^{N} (y_k - \hat{y}_k)^2 = \sum_{k=1}^{N} r_k^2 = \boldsymbol{r}^T \boldsymbol{r}. \tag{2.19}$$

Because of $\boldsymbol{X}^T \boldsymbol{X}\hat{\boldsymbol{\alpha}} = \boldsymbol{X}^T \boldsymbol{y}$, this last equation can be rewritten

$$SS_E = \boldsymbol{y}^T \boldsymbol{y} - \hat{\boldsymbol{\alpha}}^T \boldsymbol{X}^T \boldsymbol{y}. \tag{2.20}$$

Equation (2.20) is called the *error* or **residual sum of squares** (**RSS**).

2.3.2 Regression and Classification

Regression analysis is a mathematical and statistical technique which is very useful for many types of problems in engineering, science and also finance analyzing the relationship between dependent variable and one or more independent variables. Regression analysis is widely used for prediction and estimation and most commonly estimates the conditional expectation of the dependent variable given the independent variables [85]. There exist many regression methods such as Linear Regression (LR), Logit Regression, Nonlinear Regression, Generalized Linear Models, Ridge Regression and Nonparametric Regression. We explained the linear regression model in Sect. 2.3.1. In that part, we gave the least-squares method to estimate the parameters of multiple linear regression model. The present part starts with Logit Regression.

2.3.2.1 Logit Regression Models

Multivariate linear regression cannot be used to approximate categorical dependent variables, while it can be adequately used to investigate the relationship between a continuous (interval-scale) dependent variable, such as income or examination score. For that reason, instead of LR, Logit Regression is useful, especially, to model socio-economic variables [128]. It is commonly employed, especially, in GPLM, to predict sovereign debt and defaults when the dependent variable is binary, such as ‘default’ or ‘nondefault’. Since binary values (proportions) are bounded by 0 and 1, in logit regression, dependent variables do not show normal distribution properties. However, it can be assumed as Binomial distribution and, because of a variance of $\mu \cdot (1 - \mu)/n$ and a mean of μ, it is assumed as a special case of Binomial distribution: *Bernoulli distribution*, where μ is the mean and also the probability of an event occurring [128]. In this method, the maximum-likelihood estimation is used after logit transformation to the dependent variable, using the formula:

$$E(Y|\boldsymbol{x}) = P(Y = 1|\boldsymbol{x}) = H(\boldsymbol{x}^T\boldsymbol{\alpha}) = \frac{1}{1 + \exp(-\boldsymbol{x}^T\boldsymbol{\alpha})} = \mu, \tag{2.21}$$

where $\boldsymbol{x} \in \mathbb{R}^n$. Here, H is inverse link function (the cumulative distribution function), $\boldsymbol{\alpha}$ is the unknown parameter vector of the model, μ is the probability of the dependent variable to take value ‘1’ [58]. To estimate the unknown parameter vector $\boldsymbol{\alpha}$, a likelihood function is needed using the Bernoulli assumption:

$$L(\boldsymbol{\alpha}) = \prod_{k=1}^{N} \pi(\boldsymbol{x}_k; \boldsymbol{\alpha})^{y_k} (1 - \pi(\boldsymbol{x}_k; \boldsymbol{\alpha}))^{1-y_k}, \tag{2.22}$$

where $\pi(\boldsymbol{x}_k; \boldsymbol{\alpha})$ is the probability of each observation taking the value ‘1’ as dependent variable with independent variable vector $\boldsymbol{x}_j$. To facilitate the maximization of the likelihood function, the natural algorithm is applied [58]:

$$l(\boldsymbol{\alpha}) = \sum_{k=1}^{N} (y_k \ln(\pi_k(\boldsymbol{x}_k; \boldsymbol{\alpha})) + (1 - y_k) \ln(1 - \pi_k(\boldsymbol{x}_k; \boldsymbol{\alpha}))). \tag{2.23}$$

The unknown parameter vector $\hat{\boldsymbol{\alpha}}$ is obtained by solving the following equation:

$$\nabla L(\hat{\boldsymbol{\alpha}}) \left(:= \frac{\partial \ln L}{\partial \boldsymbol{\alpha}}(\hat{\boldsymbol{\alpha}}) \right) = \boldsymbol{0}. \tag{2.24}$$

To optimize the solution, iterative optimization methods, such as Newton-Raphson type method, can be used.

2.3.2.2 Nonlinear Regression Models

If there is at least one nonlinearly involved parameter in a model, this model is called as a *Nonlinear Model*. This means that in a nonlinear model at least one derivative with respect to a parameter must include that parameter. Some examples for nonlinear regression models are given as follows [119]:

$$\begin{aligned} Y &= e^{ax+bx^2} + \varepsilon, \\ Y &= ax + e^{-bx} + \varepsilon. \end{aligned} \tag{2.25}$$

Some examples for nonlinear functions are: exponential functions, logarithmic functions, trigonometric functions, power functions, Gaussian function and Lorentzian curves. Some functions, such as the exponential or logarithmic functions are assumed to be linear because they can be transformed. Here, when transformed, standard linear regression may be performed but should be employed with caution [119]. Those models which define the growth behavior over time are used in many areas. In the field of population biology, growth occurs in organisms, plants, animals, etc. [113]. The type of model which is needed in a specific situation relies on the type of growth that occurs.

In the nonlinear case, parameter estimates can also be constructed by the method of LS like in linear regression. Minimization of the RSS yields normal equations which can be nonlinear in the parameters. It is not always possible to solve nonlinear equations exactly. For this reason, the next alternative is to obtain approximate analytic solutions by using iterative procedures. For this approximate solution, three main methods are [112]:

(*a*) Linearization method,
(*b*) Steepest-Descent method, and
(*c*) Levenberg-Marquardt's method.

The *linearization method* applies the results of least-squares estimation theory in a succession of stages, but neither this method nor the steepest descent method is ideal. The linearization method converges very rapidly provided the vicinity of the true parameter values are reached. However, if initial trial values are too far removed, convergence may not occur at all, whereas the *steepest-descent method* is able to converge on true parameter values even though initial trial values are far from the true parameter values [112]. However, this convergence tends to the rather slow at the later stages of the iterative process.

The most widely applied technique of computing nonlinear LS estimators is *Levenberg-Marquardt's* method. This method presents a compromise between the other two methods and combines successfully the best features of both and avoids their serious disadvantages. It is good in the sense that it almost always converges and does not "slow down" at the latter part of the iterative process. The system is given by [112][1]

$$Y_k = f(\boldsymbol{X}_k, \boldsymbol{\alpha}) + \varepsilon_k \ \ (k = 1, 2, \ldots, N). \tag{2.26}$$

Here, Y_k is the kth observation of the dependent variable, $\boldsymbol{X}_k$ is the input part of the kth observation: $\boldsymbol{X}_k = (X_{k,1}, X_{k,2}, \ldots, X_{k,p})^T$ regarded as a random vector, $\boldsymbol{\alpha} = (\alpha_1, \alpha_2, \ldots, \alpha_p)^T$ consists of the parameters, and ε_k is the noise variable. Let the noise terms be independent and follow an $N(\alpha, \sigma^2)$ distribution. Inserting the data $(\boldsymbol{x}_k, y_k)$ $(k = 1, 2, \ldots, N)$, the residual sum of squares is given by [112]:

$$S(\boldsymbol{\alpha}) = \sum_{k=1}^{N} (y_k - f(\boldsymbol{x}_k, \boldsymbol{\alpha}))^2, \tag{2.27}$$

where $\boldsymbol{\alpha}_0 = (\alpha_{0,1}, \alpha_{0,2}, \ldots, \alpha_{0,p})^T$ is the vector of initial parameter values. The algorithm for constructing successive estimates is represented as follows:

$$(\boldsymbol{H} + \tau \boldsymbol{I})(\boldsymbol{\alpha}_0 - \boldsymbol{\alpha}_1) = \boldsymbol{g}, \tag{2.28}$$

where

$$\boldsymbol{g} = \nabla S(\boldsymbol{\alpha}_0), \ \boldsymbol{H} = \nabla^2 S(\boldsymbol{\alpha}_0). \tag{2.29}$$

Here, $\boldsymbol{I}$ is the identity matrix and τ is a suitable multiplier.

2.3.2.3 Generalized Partial Linear Models

A particular semiparametric model class of interest are the *Generalized Partial Linear Models* (*GPLMs*); they extend the Generalized Linear Models (GLMs) [85] in that the usual parametric terms are enlarged by a nonparametric component. GPLMs do not force data into any unnatural scale and so, they allow to construct a bipartite model with linear and nonlinear parts. If the normality and constant variance assumptions are not satisfied, then this approach can be applied [57].

By using a link function, GPLM makes it possible to search linear and nonlinear relationships between the mean of the response variable and the linear combination of the explanatory variables [57]. The mean value of a dependent variable rely on a linear predictor through a nonlinear link function and allows the response variable Y. For, the ease of exposition, we consider Y to follow general model that does not depend on some observation number k. In fact, the probability distribution is

[1]As we use many mathematical symbols in this book, we have a slight abuse of double use of the symbol $\boldsymbol{X}$, namely, as a vector of random input variables and as a design matrix, respectively, which should not lead to any confusion.

assumed to be any member of an exponential family of distributions Generally, a GPLM has the following form [90]:

$$E(Y|\boldsymbol{X},\boldsymbol{T}) = G(\boldsymbol{X}^T\boldsymbol{\alpha} + \gamma(\boldsymbol{T})). \tag{2.30}$$

When we use a link function $G = H^{(-1)}$, which links the mean of the dependent variable to the predictor variables, GPLM, including both parametric and nonparametric models, can be considered as an additive semiparametric model:

$$H(\mu) = \nu(\boldsymbol{X},\boldsymbol{T}) = \boldsymbol{X}^T\boldsymbol{\alpha} + \gamma(\boldsymbol{T}) = \sum_{j=1}^{p} X_j\alpha_j + \gamma(\boldsymbol{T}). \tag{2.31}$$

Here, the vectors $\boldsymbol{X}$ and $\boldsymbol{T}$ represent our decomposition of variables. While $\boldsymbol{X}$ denotes an m-variate vector of linear variables, $\boldsymbol{T}$ denotes a q-variate vector of nonlinear variables within a nonparametric model to be estimated. Furthermore, $\boldsymbol{\alpha} = (\alpha_1, \alpha_2, \ldots, \alpha_p)^T$ is the coefficient vector of $\boldsymbol{X}$ estimated by a linear (logit in our study) regression model and $\gamma(\cdot)$ is a smooth function estimated by the nonparametric model [90].

2.3.2.4 Nonparametric Regression

Nonparametric regression analysis traces the dependence of a response variable, Y_k, on one or several predictors, $x_{k,j}$ ($j = 1, 2, \ldots, p$; $k = 1, 2, \ldots, N$), without specifying in advance the function which relates the predictors to the response [45]:

$$E(Y_k) = f(x_{k,1}, x_{k,2}, \ldots, x_{k,p}) \ \ (= f(\boldsymbol{x}_k)). \tag{2.32}$$

For the sake of a compact notation, here, we write $E(Y_k)$ for the conditional expectation $E(Y_k|x_{k,1}, x_{k,2}, \ldots, x_{k,p})$. It is supposed that the conditional variance of Y_k, $\mathrm{Var}(Y_k|x_{k,1}, x_{k,2}, \ldots, x_{k,p})$ is a constant, and that the conditional distribution of Y_k is normal.

Nonparametric regression is differentiated from linear regression, in which the function relating the mean of Y_k to the x_{kj} is linear in the parameters [45]:

$$E(Y_k) = \alpha_0 + \alpha_1 x_{k,1} + \alpha_2 x_{k,2} + \ldots + \alpha_p x_{k,p}, \tag{2.33}$$

and from traditional nonlinear regression, in which the function relating the mean of Y to the x_i, though nonlinear in its parameters, is specified clearly,

$$E(Y_k) = f(x_{k,1}, x_{k,2}, \ldots, x_{k,p}; \alpha_1, \alpha_2, \ldots, \alpha_p)\,(= f(\boldsymbol{x}_k, \boldsymbol{\alpha})). \tag{2.34}$$

The easiest use of nonparametric regression consists in smoothing scatterplots. Three splines widely applied methods of nonparametric regression are kernel estimation, local-polynomial regression that is a generalization of kernel estimation, and smoothing [45]. The generalization of nonparametric regression to many predictors is mathematically straightforward. However, it is often problematic in practice.

(*i*) Multivariate data are affected by the so-called *curse of dimensionality*: Multi-dimensional spaces grow exponentially sparser with the number of dimensions, requiring very large samples to estimate nonparametric regression models with several predictors [45].

(*ii*) It is difficult to visualize a regression surface in more than three dimensions (i.e., for more than two predictors) though slicing the surface may be of some help. **Additive regression models** are an alternative to unconstrained nonparametric regression with many predictors. This regression model has the following form [45]:

$$E(Y_k) = \gamma + f_1(x_{k,1}) + f_2(x_{k,2}) + \ldots + f_p(x_{k,p}) \tag{2.35}$$

($k = 1, 2, \ldots, N$). Here, f_j are smooth partial-regression functions, estimated with smoothing splines or by local regression. An ***Additive Model*** (***AM***) can be extended in two directions:

1. To include interactions among specific predictors; for instance,

$$E(Y_k) = \gamma + f_1(x_{k,1}) + f_{23}(x_{k,2}, x_{k,3}), \tag{2.36}$$

which is not as general as the unseparated model $E(Y_k) = \gamma + f(x_{k,1}, x_{k,2}, x_{k,3})$.

2. To include linear terms, as in the model

$$E(Y_k) = \gamma + \alpha_1 x_{k,1} + f_2(x_{k,2}), \tag{2.37}$$

semiparametric models are useful for containing dummy regressors or other contrasts derived from categorical predictors. There exist some other models such as projection-pursuit regression, *Classification and Regression Trees* (*CART*) and *Multivariate Adaptive Regression Spline MARS*. In MARS, functions are of a multiplicative nature and nonsmooth. A main issue in nonparametric regression is the selection of smoothing parameters such as the span in kernel and local polynomial regression, the roughness penalty in smoothing-spline regression or equivalent degrees of freedom for any of those [45]. The statistical balance is between variance and bias, and some methods such as Cross-Validation (CV) aim to choose smoothing parameters to minimize estimated mean-square error, e.g., the sum of squared bias and variance.

2.3.3 Multivariate Adaptive Regression Splines

MARS introduced by Friedman in 1991 [46] may be presented as an extension of linear models that "automatically" models nonlinearities and interactions. It generates a multivariate-additive (multiplicative) model in a two-stage process which consists of forward and backward stage. In the *forward* stage, MARS finds *basis functions* (*BFs*) that are added to the model by a fast searching algorithm and constructs a possibly large model that overfits the dataset. The process stops when the model reaches the maximum number of BFs. However, this model at the same time contains BFs which contribute most and least to the overall performance. Thus, this forward model is quite complex and includes many incorrect terms. In the *backward* stage, the overfit model is pruned to decrease the complexity while supporting the overall performance with respect to the fit to the data. In that stage, the BFs which contribute smallest to the increase in the residual sum of squares are removed from the model at each stage and, eventually, an optimally estimated model is generated [46, 58]. MARS uses expansions of piecewise linear BFs created by dataset. The BFs, $[x-\varphi)]_+$ and $[x-\varphi]_-$, have the following form [58]:

$$[x-\varphi]_+ = \begin{cases} x-\varphi, & \text{if } x>\varphi \\ 0, & \text{otherwise} \end{cases}, \quad [x-\varphi]_- = \begin{cases} \varphi-x, & \text{if } x<\varphi \\ 0, & \text{otherwise} \end{cases}, \tag{2.38}$$

where φ is a univariate knot obtained from the dataset. These two functions are called *truncated linear functions*. Each function is piecewise linear, with a knot at the value φ, and both function together are called a *reflected pair*. The aim is to construct reflected pairs for each input x_j $(j = 1,2,\ldots,q)$ with q-dimensional knots $\boldsymbol{\varphi}_k = (\varphi_{k,1},\varphi_{k,2},\ldots,\varphi_{k,q})^T$ at each observed value $x_{k,j}$ $(k = 1,2,\ldots,N)$. Thus, the collection of BFs is written by a set of S, defined as

$$S := \left\{ \left[x_j-\varphi\right]_+, \left[x_j-\varphi\right]_- \mid \varphi \in \{x_{1,j},x_{2,j},\ldots,x_{N,j}\}, j=1,2,\ldots,q \right\}, \tag{2.39}$$

where N is the number of observations and q is the dimension of the input space. There are $2Np$ BFs if all of the input values are distinct. In *the forward stage* of MARS, the model that fits the data is built by using BFs from the set S and their products.

Note. From now on we confine ourselves to a generic response Y and a generic noise ε, which do not depend on the particular observation number k.

So, the model has the form

$$Y = \alpha_0 + \sum_{m=1}^{M} \alpha_m \vartheta_m(\boldsymbol{x}^m) + \varepsilon, \tag{2.40}$$

with an underlying vector $\boldsymbol{x} = (x_1, x_2, \ldots, x_q)^T$. Here, ε is uncorrelated random error term that is supposed to have a normal distribution with zero mean and finite variance, M is the number of BFs in the current model. Moreover, $\vartheta_m(\boldsymbol{x}^m)$ are BFs from the set S in Eq. (2.39) or *multivariate* products of two or more such functions, $\boldsymbol{x}^m$ is a subvector of $\boldsymbol{x}$ that contributes to the the function ϑ_m, and α_m are the unknown coefficients for the constant 1 ($m = 0$) or for the mth BF. Given the observations represented by the data $(\boldsymbol{x}_k, y_k)$ ($k = 1, 2, \ldots, N$), the form of the mth BF is as follows [58]:

$$\vartheta_m(\boldsymbol{x}^m) := \prod_{j=1}^{K_m} [s_{jm} \cdot (x_{v(j,m)} - \varphi_{v(j,m)})]_+. \tag{2.41}$$

Here, K_m is the number of truncated linear functions multiplied in the mth BF, $x_{v(j,m)}$ is the input variable corresponding to the jth truncated linear function in the mth BF, $\varphi_{v(j,m)}$ is the knot value corresponding to the variable $x_{v(j,m)}$ and $s_{jm} = \pm 1$.

To generate the model, the MARS forward stepwise algorithm starts with the constant function $T_0(\boldsymbol{x}^0) = 1$ to estimate α_0, and all functions in the set S are candidate functions. Possible forms of the BFs $\vartheta_m(\boldsymbol{x}^m)$ are 1, x_n, $[x_n - \varphi_i]_+$, $x_n x_l$, $[x_n - \varphi_i]_+ x_l$ and $[x_n - \varphi_i]_+ [x_l - \varphi_j]_+$ [68, 122]. For each BF, input variables cannot be the same in the MARS algorithm. Therefore, the BFs above use different input variables, x_n and x_l, and their knots, φ_i and φ_j. At each stage, all products of a function $\vartheta_m(\boldsymbol{x}^m)$ in the model set are regarded as a new BF and this term is added to the model set. That term which produces the largest decrease in the training error contains the following form [58]:

$$\alpha_{M+1}\vartheta_m(\boldsymbol{x}^m) \cdot [x_j - \varphi]_+ + \alpha_{M+2}\vartheta_m(\boldsymbol{x}^m) \cdot [\varphi - x_j]_+.$$

Here, α_{M+1} and α_{M+2} are coefficients and they are determined by least-squares estimation, along with all other $M+1$ coefficients in the model. Then, the "winning" products are added to the model and the process stops as soon as the model set reaches some present maximum number of terms. At the end of this forward stepwise process, a large model of the form is obtained. This model does typically *overfit* the data, and so a backward deletion procedure is applied.

The *backward stepwise algorithm* removes the terms that contribute the smallest increase in the residual squared error from the model at each stage, and this iterative procedure continues until an optimal number of effective terms are present in the final model [46]. So, an estimated best model $\hat{f}_\beta$ of each number of terms β is produced at the end of this process. In the MARS model, *generalized cross-validation (GCV)* is used to find the optimal number of terms β. It also shows the lack of fit when using MARS. The GCV criterion defined by Friedman [46] is defined as follows:

$$LOF(\hat{f}_\beta) = GCV(\alpha) := \frac{\sum_{k=1}^{N}(y_k - \hat{f}_\beta(\boldsymbol{x}_k))^2}{(1 - M(\beta)/N)^2}. \tag{2.42}$$

Here, $M(\beta)$ is the effective number of parameters in the model, and N is the number of sample observations, i.e., of the data [58].

2.3.4 Tikhonov Regularization

A problem is defined as *ill-posed problem* if a solution is not existing or not unique or if it is not stable under perturbation on data—that is, if an arbitrarily small perturbation of the data can cause an arbitrarily large perturbation of the solution [53]. *Tikhonov Regularization* (*TR*) is the most common and well-known form to make these problems regular and stable. For statistics, it is also known as *ridge regression*.

TR method searches the regularized solution as a minimizer of a weighted combination of the residual norm and a side constraint. The regularization parameter controls the weight given to the minimization of the side constraint. Therefore, the quality of the regularized solution is controlled by the regularization parameter. An optimal regularization parameter should fairly balance between the size of the residual error and the stabilizing of the approximate solution [67]. A suitable value of the regularization parameter is considered and computed when the norm of the error in the data or the norm of the solution of the error-free problem are available.

The regularization parameter brings the optimal rate of convergence for the approximations, which are generated by the application of TR to ill-posed equations [91]. However, when we derive rates of convergence, we must make assumptions about the nature of the stabilization (i.e., the choice of the semi norm in the TR) and the regularity imposed on the solution. In fact, there is a **trade-off** between stabilization and regularity in terms of the rate of convergence.

The *L-curve criterion* is a practical method for choosing regularization parameter when data are noisy. The method is based on the plot of the norm of the regularized solution versus the norm of the corresponding residual [54]. The idea of the L-curve criterion is to select a regularization parameter related to the characteristic L-shaped corner of the graph. The corner shows where the curve is closest to the origin and where the curvature is maximal. However, when it is plotted in a linear scale, it is difficult to inspect the features of the L-curve because of the large range of values for the two norms. The features become easier to inspect when the curve is plotted in the double logarithmic scale [54]. Therefore, in many cases it is better to analyze the L-curve in the log-log scale.

For TR, the L-curve is important in the analysis of discrete ill-posed problems. The L-curve shows how the regularized solution changes as the regularization parameter changes. The corner of the L-curve corresponds to a good balance between the minimization of the sizes, and the corresponding regularization parameter is a good one, because a distinct L-shaped corner of the L-curve is located exactly where the solution changes, from being dominated by the regularization errors to being dominated by right-hand side errors [67].

Tikhonov solution can be expressed easily in terms of the *Singular Value Decomposition* (*SVD*) of the coefficient matrix $\boldsymbol{A}$ of regarded linear systems of equations

$$\boldsymbol{A}\boldsymbol{x} = \boldsymbol{b}, \tag{2.43}$$

where $\boldsymbol{A}$ is an ill-conditioned $(N \times n)$-matrix. The standard approach to approximately solve this system of equations is known as (*linear*) LS estimation. It seeks to minimize the residual $\|\boldsymbol{b} - \boldsymbol{A}\boldsymbol{x}\|_2^2$. There can be infinitely many solutions for a general linear LS problem. If it is considered that the data contain noise, in that situation, the data points cannot be fitted exactly because of noise. It becomes evident that there can be many solutions, which can adequately fit the data in the sense that the Euclidean distance $\|\boldsymbol{b} - \boldsymbol{A}\boldsymbol{x}\|_2$ is smallest. The *discrepancy principle* [5] can be used to regularize the solution of a discrete ill-posed problem based on the assumption that a reasonable level for $c = \|\boldsymbol{b} - \boldsymbol{A}\boldsymbol{x}\|_2$ is known.

Different kinds of TR are represented as minimization problems. Under the discrepancy principle, all solutions with $\|\boldsymbol{b} - \boldsymbol{A}\boldsymbol{x}\|_2 \leq c$ are considered, and we select the one that minimizes the norm of $\boldsymbol{x}$:

$$\begin{aligned} &\underset{\boldsymbol{x}}{\text{minimize}}\ \|\boldsymbol{x}\|_2\,, \\ &\text{subject to}\ \|\boldsymbol{b} - \boldsymbol{A}\boldsymbol{x}\|_2 \leq c, \end{aligned} \tag{2.44}$$

or we minimize the norm of residual vector under some tolerance with respect to the norm of $\boldsymbol{x}$:

$$\begin{aligned} &\underset{\boldsymbol{x}}{\text{minimize}}\ \|\boldsymbol{b} - \boldsymbol{A}\boldsymbol{x}\|_2\,, \\ &\text{subject to}\ \|\boldsymbol{x}\|_2 \leq d. \end{aligned} \tag{2.45}$$

In the first optimization problem in Eq. (2.44), any important nonzero feature that appears in the regularized solution increases $\|\boldsymbol{x}\|_2$. However, these features exist in the solution because they are necessary to fit the data. Therefore, the minimum of $\|\boldsymbol{x}\|_2$ guarantee that unimportant features should be removed in the regularized solution. As c increases, the set of feasible models expands, and the minimum value of $\|\boldsymbol{x}\|_2$ decreases.

In the second optimization problem in Eq. (2.45), it is wanted to choose the minimum norm solution among those parameter vectors, which adequately fit the data, because any important nonzero feature that appears in the regularized solution must not be ignored to fit the data, and unimportant data must be removed by the regularization. As d decreases, the set of all feasible solutions becomes smaller, and the minimum value of increases.

There is also a third option which is considered a dampened LS problem:

$$\underset{\boldsymbol{x}}{\text{minimize}}\ \|\boldsymbol{b} - \boldsymbol{A}\boldsymbol{x}\|_2^2 + \lambda_2\,\|\boldsymbol{x}\|_2^2\,, \tag{2.46}$$

arising when the method of Lagrange multipliers is applied to problem in Eq. (2.45). Here, λ is the tradeoff parameter between the first and the second part. The problems in Eqs. (2.44)–(2.46) have the same solution for some appropriate choice of the values α, β and λ [5].

To solve different kinds of TR problem discussed above, we use Singular Value Decomposition (SVD) to have a solution that minimizes the objective function including $\|\boldsymbol{x}\|_2$. However, in many cases, it is preferred to achieve a solution that minimizes some other measure of $\boldsymbol{x}$, such as the norm of first- or second-order derivatives. These derivatives are, in an approximative sense, given by first- or second-order difference quotients of $\boldsymbol{x}$ which is considered as a function that is evaluated at the discrete points k and $k+1$. These difference quotients approximate first- and second-order derivates; altogether, they are comprised by products $\boldsymbol{Lx}$ of $\boldsymbol{x}$ with matrices $\boldsymbol{L}$. These matrices represent the discrete differential operators of first- and second-order, respectively [5]. Hereby, the optimization problem is of the following form:

$$\underset{\boldsymbol{x}}{\text{minimize}} \; \|\boldsymbol{b} - \boldsymbol{Ax}\|_2^2 + \lambda_2 \|\boldsymbol{Lx}\|_2^2 \,. \tag{2.47}$$

The optimization problem of Eq. (2.47) turns into the optimization problem of Eq. (2.46) when $\boldsymbol{L} = \boldsymbol{I}$. Then, it is called *zeroth order* TR, which is a special case of Eq. (2.47). Generally, Eq. (2.47) consists of high order TR problems. Although zeroth-order TR is solved based on SVD, to one concerned with higher-order TR, *generalized SVD* is used. In many situations, to obtain a solution which minimizes some other measure $\boldsymbol{x}$, the norm of the first- or second-order derivatives is preferred [55].

Chapter 3
New Robust Analytic Tools

In the previous chapter, we mentioned about some mathematical methods that are used in this book. In present chapter, we introduce our robust tools, R(C)MARS and R(C)GPLM, in theory and method.

3.1 Robust (Conic) Multivariate Adaptive Regression Splines

3.1.1 Introduction

(C)MARS models depend on parameters, and small perturbations in the data may result in different parameter estimates, and hence, may bring about unstable solutions. Indeed, measurement error that affects the independent variables in regression models is a common problem in many scientific areas. It is well known that the implications of ignoring measurement errors in inferential procedures may be substantial, often resulting in unreliable results [7, 28]. In order to reduce the estimation variance while keeping the efficiency as high as possible, we robustified the (C)MARS method by using approaches such as scenario optimization and robust counterpart. We are interested in the multicriteria **tradeoff** (antagonism) between ***accuracy*** and ***robustness***. In the line of our research [96–99, 101, 102], robustness has become, in some sense, an extension of stability or regularity. Stability also means a small complexity of the model, or: a *small variance* of the estimation.

Through ***RCMARS*** we are also permitted to involve uncertainty in the input variables into regression and classification modeling; that uncertainty is typical for real-world challenges, too. In fact, in RCMARS, we have implied uncertainty in both input and output variables. This means that in RCMARS, there is a double way of robustification: (*a*) The regularization (stabilization) in integral form that expresses itself in the involvement of the (squared) first- and second-order partial

A. Özmen, *Robust Optimization of Spline Models and Complex Regulatory Networks*, Contributions to Management Science,
DOI 10.1007/978-3-319-30800-5_3

derivatives of the multivariate basis functions; after discretization of the integrals, we arrive at a TR problem [5] with first- and second-order complexity terms. This TR problem is turned into a CQP problem [127, 136]. (*b*) The robustification is performed with respect to the input variables and output or response variable, all of them with their own uncertainty sets.

In (*a*), via those first- and second-order terms, we aim at a flat model and a one where high *energy* in the model (curvature) is penalized so that we could speak of a 'dampened' or 'tamed' model, respectively. This also means that in RCMARS, we have, in addition to the robustification, an additional support of the robustification agenda, an *increase* of robustness, whereby that support is of a fine-tuned kind of character which is parametric through the bounds of complexity in the CQP program. For our RCMARS, we conduct a penalization in the form of TR and study it as a RCQP problem in order to achieve a reduction in the complexity of the regression method MARS that especially means sensitivity with respect to ***noise*** in the data.

In contrast, in our ***RMARS***, we only have the robustification step (*b*), whereas the aforementioned fine-tuning option (*a*) dropped. At the first glance, this seems to be a qualitative loss. However, RCMARS is leading to a very large computational effort, and parametric studies which are enabled by part (*a*) do even increase those computational costs. It belonged to the main ideas of MARS and CMARS to have (*I*) a 'doable' methods, even with an effect in (*II*) a variance reduction for the estimated model. Here, we pay tribute to these important aims (*I*) and (II), in the form of our simplified and 'handy' alternative of RCMARS, called RMARS.

Briefly, (C)MARS are robustified through the robust optimization approach, which is some rigorous kind of regularization in the input and output domain. We have some generalization effect now in the part of $\left\| \boldsymbol{y} - \boldsymbol{\vartheta}(\boldsymbol{b})\boldsymbol{\alpha} \right\|_2^2$, when we conduct our R(C)MARS for both input and output variables by including uncertainty, via RO [9, 10, 13, 14]. However, in RCMARS, we need not to make any change in the additional integration term on the complexity, or "energy". By introducing R(C)MARS, we aim to decrease the **estimation variance**.

3.1.2 The Procedure

The MARS [58] method supposes the following general model

$$Y = f(\boldsymbol{X}) + \varepsilon, \tag{3.1}$$

where Y is the response variable; $\boldsymbol{X} = (X_1, X_2, \ldots, X_q)^T$ is a vector of predictor variables; ε is an additive stochastic component with zero mean and finite variance. It aims to build reflected pairs for each input X_j ($j = 1, 2, \ldots, q$) with q-dimensional knots $\boldsymbol{\varphi} = (\varphi_1, \varphi_2, \ldots, \varphi_q)^T$ at or just nearby each of the input data vectors $\boldsymbol{x}_k = (x_{k,1}, x_{k,2}, \ldots, x_{k,q})^T$ ($k = 1, 2, \ldots, N$), where q and N represent the number of predictors and observations, respectively. For this purpose, first, the set of BFs is formed by an intensive but a fast search procedure as follows:

$$S := \left\{[X_j - \varphi]_+, [X_j - \varphi]_- \mid \varphi \in \{x_{1,j}, x_{2,j}, \ldots, x_{N,j}\}, j = 1, 2, \ldots, q\right\}. \tag{3.2}$$

Each function in S, a *reflected pair*, is piecewise linear with a knot value, φ. Then, Y becomes

$$Y = \alpha_0 + \sum_{m=1}^{M} \alpha_m \vartheta_m(\boldsymbol{x}^m) + \varepsilon. \tag{3.3}$$

Here, ϑ_m $(m = 1, 2, \ldots, M)$ is a basis function (BF) from S or products of two or more such functions: α_m is the unknown coefficient associated with the mth BF $(m = 1, 2, \ldots, M)$, where m equals zero for the constant one and M is the number of BFs. When the data is represented by $(\boldsymbol{x}_k, y_k)$ $(k = 1, 2, \ldots, N)$, the mth BF takes the following form

$$\vartheta_m(\boldsymbol{x}^m) := \prod_{j=1}^{K_m} [s_{jm} \cdot (x_{v(j,m)} - \varphi_{v(j,m)}]_+. \tag{3.4}$$

In the CMARS method, to estimate Y in Eq. (3.1), instead of the backward stepwise algorithm of MARS, an alternative method [127] is utilized, in which penalty terms are used in addition to the least-squares estimation (LSE) to control the lack-of-fit with regard to the complexity and stability. Consequently, the *Penalized Residual Sum of Square* (*PRSS*) with $M_{\max}$ BFs is formed as

$$PRSS := \sum_{k=1}^{N} (y_k - \boldsymbol{\alpha}^T \boldsymbol{\vartheta}(\boldsymbol{b}_k))^2 + \sum_{m=1}^{M_{\max}} \phi_m \sum_{\substack{|\boldsymbol{\theta}|=1 \\ \boldsymbol{\theta}^T=(\theta_1,\theta_2)}}^{2} \sum_{\substack{r<s \\ r,s \in V(m)}} \int_{Q^m} \alpha_m^2 [D_{r,s}^{\boldsymbol{\theta}} \vartheta_m(\boldsymbol{t}^m)]^2 d\boldsymbol{t}^m, \tag{3.5}$$

where $\boldsymbol{\vartheta}(\boldsymbol{b}_k) := (1, \vartheta_1(\boldsymbol{x}_k^1), \ldots, \vartheta_m(\boldsymbol{x}_k^{M_{\max}}))$; $V(m) := \{v(k,m) \mid j = 1, 2, \ldots, K_m\}$ is the variable set associated with the mth BF called ϑ_m; $\boldsymbol{t}^m = (t_{m_1}, \ldots, t_{m_{K_m}})^T$ represents the vector of variables that contribute to the mth BF, ϑ_m; $\boldsymbol{\alpha}$ is an $((M_{\max} + 1) \times 1)$- parameter vector to be estimated using the data points; $\phi_m \geq 0$ are the *penalty parameters* $(m = 1, 2, \ldots, M_{\max})$. Moreover, Q^m is some appropriately large K_m-dimensional parallelpipe where the integration occurs. Furthermore,

$$D_{r,s}^{\boldsymbol{\theta}} \vartheta_m(\boldsymbol{t}^m) = (\partial^{|\boldsymbol{\theta}|} \vartheta_m) / (\partial^{\theta_1} t_r^m \partial^{\theta_2} t_s^m) \boldsymbol{t}^m$$

expresses the first- or second-order derivatives, where $\boldsymbol{\theta} = (\theta_1, \theta_2)^T, |\boldsymbol{\theta}| := \theta_1 + \theta_2$ and $\theta_1, \theta_2 \in \{0, 1\}$. Since it is not easy to evaluate the multi-dimensional integrals in Eq. (3.5), a discretization is applied to approximate the integral $\int_{Q^m} \alpha_m^2 [D_{r,s}^{\theta} \vartheta_m(\boldsymbol{t}^m)]^2 d\boldsymbol{t}^m$ (cf. [96, 127] for more details). Therefore, the approximation of PRSS in Eq. (3.5) can be rearranged as

$$PRSS \approx \left\| \boldsymbol{y} - \boldsymbol{\vartheta}(\boldsymbol{b})\boldsymbol{\alpha} \right\|_2^2 + \phi \left\| \boldsymbol{L}\boldsymbol{\alpha} \right\|_2^2, \tag{3.6}$$

where $\boldsymbol{L}$ is an $((M_{\max} + 1) \times (M_{\max} + 1))$-diagonal matrix. Afterwards, the *PRSS* problem turns into a classical *Tikhonov Regularization* (*TR*) [5] problem if we employ only one penalty factor $\phi > 0$, $\phi = \lambda^2$ for some $\lambda \in \mathbb{R}$ instead of using different penalty parameters. So, the PRSS form in Eq. (3.6) may be formulated as a CQP [12, 35] and, using an appropriate bound $K \geq 0$, the following optimization problem can be stated [136]:

$$\begin{aligned} &\underset{w,\boldsymbol{\alpha}}{\text{minimize}} \; w \\ &\text{subject to} \left\| \boldsymbol{y} - \boldsymbol{\vartheta}(\boldsymbol{b})\boldsymbol{\alpha} \right\|_2 \leq w, \\ &\qquad\qquad \left\| \boldsymbol{L}\boldsymbol{\alpha} \right\|_2 \leq \sqrt{K}. \end{aligned} \tag{3.7}$$

Here, the choice of the parameter K has to be the outcome of a careful learning process, with the help of model-free or model-based methods [5].

Remark 1 In future studies, we go on facing the complexity of our model and trying to turn all model-free, e.g., trial-and-error, sides of our treatment, into a model-based form. In particular, we plan to reinterpret a parametric bound such a K as another state variable (unknown), including it into the objective function also. Herewith, we would still remain in our 'conic' setting of CQP. This could lead to another support and strengthening of the model-basedness of our approach and would make it even more rigorous mathematically. Modern continuous and global optimization will certainly be a key-technology for this. We can also diversify our optimization by differentiating between different values of the penalty parameters. This would lead to further *control* variables.

In R(C)MARS, we assume that the input and output variables of our model are *random variables* all. They lead us to *uncertainty sets*; those are assumed to contain *confidence intervals* (*CIs*) (we refer to [98, 101] for more details). For CMARS, the large model that has the maximum number of BFs, $M_{\max}$, is created by Salford MARS® [83]. The following general model represents the relation between both the *random* input variables and the response, itself being affected with noise:

$$Y = f(\underbrace{\boldsymbol{X}}_{\text{noisy variable}}) + \varepsilon, \tag{3.8}$$

where $\boldsymbol{X} = (X_1, X_2, \ldots, X_q)^T$ is a vector of random predictor variables. The random variables X_j are assumed to be normally distributed. Here, the following general model is considered for each input X_j [97, 98]:

$$X_j = \bar{x}_j + \xi_j. \tag{3.9}$$

When considering that we have q-dimensional input data and incorporate a "*perturbation*" (*uncertainty*) into the input data, each input data vector $\boldsymbol{x}_k = (x_{k,1}, x_{k,2}, \ldots, x_{k,q})^T$ is represented as $\breve{\boldsymbol{x}}_k = (\breve{x}_{k,1}, \breve{x}_{k,2}, \ldots, \breve{x}_{k,q})^T$, including the perturbation $\boldsymbol{\Delta}_k = (\Delta_{k,1}, \Delta_{k,2}, \ldots, \Delta_{k,q})^T$ $(k = 1, 2, \ldots, N)$. Since, in each coordinate, some values $x_{k,j}$ can be outlier, but the perturbation of an outlier is not meaningful, for our problem, we, instead, refer to $\bar{x}_j = (1/N) \cdot \sum_{k=1}^{N} x_{kj}$, the mean (average) of the input vector $\boldsymbol{x}_j$, as the reference value wherever we use $\boldsymbol{x}_j$. Here, $\boldsymbol{\Delta}_k$ is a generic element of U_1, which is the uncertainty set for our input data. Herewith, our new values of piecewise linear BFs are shown in the following:

$$x_{k,j} \to \breve{x}_{k,j}; \qquad \breve{x}_{k,j} = \bar{x}_j + \Delta_{k,j}, \ \ |\Delta_{k,j}| \le \rho_{k,j} \ \ (k = 1, 2, \ldots, N; j = 1, 2, \ldots, q), \tag{3.10}$$

where $x_{k,j}$ is an noisy input value; $\breve{x}_{k,j}$ is an input value that has uncertainty; $\Delta_{k,j}$ is a perturbation of $x_{k,j}$; $\rho_{k,j}$ is the semilength of the CI for input data, and the amount of perturbation in each dimension is restricted by $\rho_{k,j}$. Similarly, when we incorporate a "*perturbation*" (*uncertainty*) into output data, our output data vector $\boldsymbol{y} = (y_1, y_2, \ldots, y_N)^T$ is stated as $\breve{\boldsymbol{y}} = (\breve{y}_1, \breve{y}_2, \ldots, \breve{y}_N)^T$ including the perturbation $\boldsymbol{\tau} = (\tau_1, \tau_2, \ldots, \tau_N)^T$. As, again, some values y can be outlier and the perturbation of an outlier is not meaningful, for our problem, we refer to $\bar{y} = (1/N) \cdot \sum_{k=1}^{N} y_k$, the average of the output vector $\boldsymbol{y}$, as the reference value wherever we refer to $\boldsymbol{y}$. Here, we restrict the vector $\boldsymbol{\tau}$ to be elements of U_2, being the uncertainty set for our output data. So, our new output values can be represented by [101]:

$$y_k \to \breve{y}_k; \qquad \breve{y}_k = \bar{y} + \tau_k, \ \ |\tau_k| \le \nu_k \ \ (k = 1, 2, \ldots, N). \tag{3.11}$$

Here, the amount of perturbation is limited by ν_k which is the semilength of the CI for the output data. In order to robustify (C)MARS, we employ some robust optimization on the BFs provided by the MARS model. MARS method constructs expansions of piecewise linear BFs; by this, it will be based on the new dataset that includes uncertainty. Aiming at the variable $\breve{x}$ we prefer the following notation for the piecewise linear BFs [58]:

$$c^+(\breve{x}, \varphi) = [\breve{x} - \varphi]_+, \quad c^-(\breve{x}, \varphi) = [\breve{x} - \varphi]_-. \tag{3.12}$$

Incorporating the uncertainty sets $U_1 \subseteq \mathbb{R}^{N \times M_{\max}}$ and $U_2 \subseteq \mathbb{R}^N$, determined below in Sect. 3.1.3, into the data $(\breve{\boldsymbol{x}}_k, \breve{y}_k)$, the multiplicative form of the mth BF can be stated as

$$\vartheta_m(\breve{\boldsymbol{x}}_k^m) = \prod_{j=1}^{K_m} [\breve{x}_{k,v(j,m)} - \varphi_{v(j,m)}]_{\pm} \quad (k = 1, 2, \ldots, N). \tag{3.13}$$

When estimating the BFs $[\breve{x}_{k,v(j,m)} - \varphi_{v(j,m)}]_\pm$ in Eq. (3.13), we can evaluate them by the following special terms of estimation [98]:

$$[\breve{x}_{k,v(j,m)} - \varphi_{v(j,m)}]_\pm \leq [x_{k,v(j,m)} - \varphi_{v(j,m)}]_\pm + [\Delta_{k,v(j,m)} + (\pm \mathrm{A}_{k,v(j,m)})]_\pm. \tag{3.14}$$

Here, $\mathrm{A}_{k,v(j,m)}$ is interpreted and employed as *control parameters*. If we consider the *risk friendly* case, we select the value of $\mathrm{A}_{k,v(j,m)}$ between 0 and the absolute value of $\mathrm{A}_{k,v(j,m)}$, i.e., $\breve{\mathrm{A}}_{k,v(j,m)} \in [0, |\mathrm{A}_{k,v(j,m)}|]$. Now, to simplify the notation, we still preserve the notion $\mathrm{A}_{k,v(j,m)}$ for $\breve{\mathrm{A}}_{k,v(j,m)}$. To estimate the values $\vartheta(\boldsymbol{x}_k)$ and $\vartheta(\breve{\boldsymbol{x}}_k)$, we can employ Eq. (3.13) in the subsequent form, where all the "+" and "-" signs belong to each other, respectively:

$$\begin{aligned}
\underbrace{\prod_{j=1}^{K_m} [\breve{x}_{k,v(j,m)} - \varphi_{v(j,m)}]_\pm}_{=:\vartheta_m(\breve{\boldsymbol{x}}_k)} \leq \underbrace{\prod_{j=1}^{K_m} [x_{k,v(j,m)} - \varphi_{v(j,m)}]_\pm}_{=:\vartheta_m(\boldsymbol{x}_k)} + \\
\sum_{A \underset{\neq}{\subseteq} \{1,\ldots,K\}} \prod_{a \in A} [x_{ka} - \tau_a]_\pm \prod_{b \in \{1,\ldots,K\}/A} [(\pm \mathrm{A}_{kb}) + \Delta_{kb}]_\pm \quad (k = 1, 2, \ldots, N).
\end{aligned} \tag{3.15}$$

Then, for each BF, the uncertainty value $|u_{km}|$ can be estimated in the subsequent way:

$$\begin{aligned}
|u_{km}| &\leq \sum_{A \underset{\neq}{\subseteq} \{1,\ldots,K\}} \prod_{a \in A} \underbrace{|x_{ka} - \tau_a|}_{\leq D_{ka}\rho_{ka}} \prod_{b \in \{1,\ldots,K\}/A} \underbrace{(| \pm \mathrm{A}_{k,v(j,m)} + \Delta_{k,v(j,m)}|)}_{\leq \gamma_{kb} + \rho_{kb}} \\
&\leq \sum_{A \underset{\neq}{\subseteq} \{1,\ldots,K\}} \prod_{a \in A} \mathrm{B}_{ka}\rho_{ka} \prod_{b \in \{1,\ldots,K\}/A} (\gamma_{kb} + \rho_{kb}) \\
&\leq \sum_{A \underset{\neq}{\subseteq} \{1,\ldots,K\}} \prod_{a \in A} \underbrace{\mathrm{B}_{ka}}_{\leq \mathrm{B}_k} \prod_{a \in A} \rho_{ka} \prod_{b \in \{1,\ldots,K\}/A} (\gamma_{kb} + \rho_{kb}) \\
&\leq \sum_{A \underset{\neq}{\subseteq} \{1,\ldots,K\}} \mathrm{B}_k^{|A|-1} \prod_{a \in A} \rho_{ka} \prod_{b \in \{1,\ldots,K\}/A} (\gamma_{kb} + \rho_{kb}),
\end{aligned} \tag{3.16}$$

where the amount of the value of $\mathrm{A}_{k,v(j,m)}$ is restricted by γ, the cardinality of the set A has been denoted through $|A|$, and B_k is also considered to be applied as a *control parameter*. The value of B_k is equal to 2 in cases without outliers, but for outliers, it will be greater than 2. For such a case, we will have to select a different value for B_k.

Now, for RCMARS, PRSS in Eq. (3.6) will have the following approximate representation:

$$PRSS \approx \left\| \breve{y} - \boldsymbol{\vartheta}(\breve{\boldsymbol{b}})\boldsymbol{\alpha} \right\|_2^2 + \phi \left\| \boldsymbol{L}\boldsymbol{\alpha} \right\|_2^2. \tag{3.17}$$

Herewith, the *PRSS* minimization problem again looks like a classical TR problem [5] with $\phi > 0$, i.e., $\phi = \lambda^2$ for some $\lambda \in \mathbb{R}$, and then, it can be coped with through CQP [12, 35]. The second (complexity) part of the PRSS approximation remains the same as it is in CMARS after we incorporate a "*perturbation*" into the real input data $\boldsymbol{x}_k$, in each dimension, and into the output data y_k, since we do not make any changes for the function in the multi-dimensional integrals.

3.1.3 Polyhedral Uncertainty and Robust Counterparts

As it is known, robustification is more successful when ellipsoidal uncertainty sets are employed, rather than polyhedral uncertainty sets. Nevertheless, using ellipsoidal uncertainty sets can increase the complexity of our optimization models [146]. We study *robust CQP* (*RCQP*)(or *robust second-order optimization problem, RSCOP*) under polyhedral uncertainty and we shall find out that it equivalently means a *standard CQP*. To analyze the robustness problem, we assume that the given model uncertainty is represented by a family of matrices $\boldsymbol{\vartheta}(\breve{\boldsymbol{x}}) = \boldsymbol{\vartheta}(\boldsymbol{x}) + \boldsymbol{U}$ and vectors $\breve{\boldsymbol{y}} = \boldsymbol{y} + \boldsymbol{v}$, where U_1, containing $\boldsymbol{U}$, and U_2, containing $\boldsymbol{v}$, are bounded sets which need to be specified first. Here, the uncertainty matrix $\boldsymbol{U} \in U_1$ and uncertainty vector $\boldsymbol{v} \in U_2$ are of the formats

$$\boldsymbol{U} = \begin{bmatrix} u_{1,1} & u_{1,2} & \cdots & u_{1,M_{\max}} \\ u_{2,1} & u_{2,2} & \cdots & u_{2,M_{\max}} \\ \vdots & \vdots & \ddots & \vdots \\ u_{N,1} & u_{N,2} & \cdots & u_{N,M_{\max}} \end{bmatrix} \text{ and } \boldsymbol{v} = \begin{bmatrix} v_1 \\ v_2 \\ \vdots \\ v_N \end{bmatrix}. \tag{3.18}$$

As we do not want to increase the overall complexity of our optimization problems, we select the uncertainty sets U_1 and U_2 of type *polyhedral* for both input and output data in our model, to study our robustness problem. Based on these sets, the *robust counterpart* of CMARS is defined as

$$\underset{\boldsymbol{\alpha}}{\text{minimize}} \max_{\substack{W \in U_1, \\ z \in U_2,}} \left\| z - \boldsymbol{W}\boldsymbol{\alpha} \right\|_2^2 + \phi \left\| \boldsymbol{L}\boldsymbol{\alpha} \right\|_2^2, \tag{3.19}$$

with some $\phi \geq 0$. Now, we can receive the *robust counterpart* of MARS if we drop the second part (complexity part) of Eq. (3.19). Here, the uncertainty set U_1 is a polytope with $2^{N \cdot M_{\max}}$ vertices $\boldsymbol{W}^1, \boldsymbol{W}^2, \ldots, \boldsymbol{W}^{2^{N \cdot M_{\max}}}$. In fact, although it is not a known singleton, it allows a representation:

$$U_1 = \left\{ \sum_{\kappa=1}^{2^{N \cdot M_{\max}}} \eta_\kappa \boldsymbol{W}^\kappa \mid \eta_\kappa \geq 0 \;\; (\kappa \in \{1, 2, \ldots, 2^{N \cdot M_{\max}}\}), \; \sum_{\kappa=1}^{2^{N \cdot M_{\max}}} \eta_\kappa = 1 \right\}, \tag{3.20}$$

i.e., $U_1 = \text{conv}\{\boldsymbol{W}^1, \boldsymbol{W}^2, \ldots, \boldsymbol{W}^{2^{N \cdot M_{\max}}}\}$ is the convex hull. Furthermore, U_2 is a polytope with 2^N vertices $z^1, z^2, \ldots, z^{2^N}$ having the form

$$U_2 = \left\{ \sum_{\mu=1}^{2^N} \psi_\mu z^\mu | \psi_\mu \geq 0 \; (\mu \in \{1, 2, \ldots, 2^N\}), \; \sum_{\mu=1}^{2^N} \psi_\mu = 1 \right\}, \tag{3.21}$$

where $U_2 = \text{conv}\{z^1, \ldots, z^{2^N}\}$ is the convex hull. Here, any uncertainty sets U_1 and U_2 can be represented as a convex combination of vertices $\boldsymbol{W}^\kappa$ $(\kappa \in \{1, \ldots, 2^{N \cdot M_{\max}}\})$ and z^μ $(\mu \in \{1, \ldots, 2^N\})$ of the polytope, respectively. The entries are found to have become intervals. Therefore, our matrix $\boldsymbol{W}$ and vector z with uncertainty are lying in the Cartesian product of intervals that are parallelpipes (see [96, 98] for more details). To give an easy illustration, the Cartesian product of intervals in general and, especially, for three entries can be represented by Fig. 3.1.

Here, we represented the matrix $\boldsymbol{W}$ as a vector with uncertainty which generates a parallelpipe. We have a $(N \times M_{\max})$-matrix $\boldsymbol{W} = (w_{kj})_{\substack{k=1,2,\ldots,N \\ j=1,2,\ldots,M_{\max}}}$ and we can write it as a vector $\boldsymbol{t} = (t_l)_{l=1,2,\ldots,N \cdot M_{\max}}$, where $t_l := u_{kj}$ with $l = k + (j-1)N$. So, our matrix $\boldsymbol{W}$ can be canonically represented as a vector $\boldsymbol{t} = (t_1, t_2, \ldots, t_{N \cdot M_{\max}})^T$ by putting the columns of $\boldsymbol{W}$ behind each other.

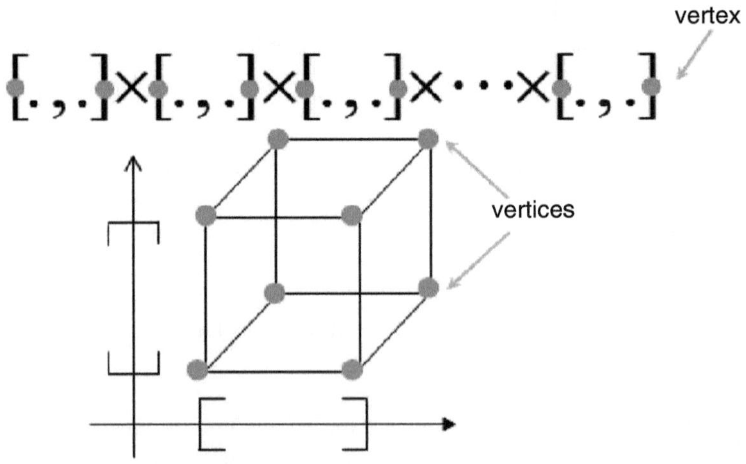

Fig. 3.1 Cartesian product of intervals for three entries [96]

Remark 2 Whenever we use polyhedral uncertainty sets, we have a drawback in practice since there may be too many vertices to handle them computationally or we might not know them exactly. That drawback comes from a very high complexity and consists in the resulting storage and and computing problems at common workstations. In fact, with polyhedral uncertainty, our matrix $\boldsymbol{W}$ represented as a vector $\boldsymbol{t}$ has a very big dimension in our applications and our computer capacity is not enough to solve our optimization problem with $\boldsymbol{W}$. Because of this, we need to discuss weak robustification case in our applications (cf. Sects. 3.1.5, 5.2.1, 6.1.7). That weak robustification encounters a data-wise robustification which refers to all the other data according to the interval midpoints ("ceteris paribus"), and it eventually addresses the worst case with respect to all the data-wise robustifications.

3.1.4 Robust Conic Quadratic Programming with Polyhedral Uncertainty

For *RCMARS* model, the optimization problem is written as follows:

$$\begin{aligned} \underset{w,\boldsymbol{\alpha}}{\text{minimize}} \quad & w \\ \text{subject to} \quad & \left\| \breve{\boldsymbol{y}} - \boldsymbol{\vartheta}(\breve{\boldsymbol{b}})\boldsymbol{\alpha} \right\|_2 \leq w, \\ & \left\| \boldsymbol{L}\boldsymbol{\alpha} \right\|_2 \leq \sqrt{K}, \end{aligned} \tag{3.22}$$

with some parameter $K \geq 0$. Via the height variable w (by an epigraph argument), recalling that U_1 and U_2 are polytopes, described by their vertices in Eqs. (3.20)–(3.21), the RCQP for our optimization problem is equivalently represented as a standard CQP in the following form:

$$\begin{aligned} \underset{w,\boldsymbol{\alpha}}{\text{minimize}} \quad & w \\ \text{subject to} \quad & \left\| \boldsymbol{W}\boldsymbol{\alpha} - \boldsymbol{z} \right\|_2 \leq w \quad \forall \underbrace{\boldsymbol{W}}_{=\sum_{\kappa=1}^{2^{N \cdot M_{\max}}} \eta_\kappa \boldsymbol{W}^\kappa} \in U_1, \quad \underbrace{\boldsymbol{z}}_{=\sum_{\mu=1}^{2^N} \psi_\mu \boldsymbol{z}^\mu} \in U_2, \\ & \left\| \boldsymbol{L}\boldsymbol{\alpha} \right\|_2 \leq \sqrt{K}. \end{aligned} \tag{3.23}$$

Here, U_1 and U_2 are polytopes which are described by their vertices as

$$U_1 = \text{conv}\{\boldsymbol{W}^1, \boldsymbol{W}^2, \ldots, \boldsymbol{W}^{2^{N \cdot M_{\max}}}\}, \quad U_2 = \text{conv}\{\boldsymbol{z}^1, \boldsymbol{z}^2, \ldots, \boldsymbol{z}^{2^N}\}. \tag{3.24}$$

Therefore, our RCQP can be equivalently stated by a standard CQP as follows:

$$\begin{aligned} &\underset{w,\boldsymbol{\alpha}}{\text{minimize}} \ w \\ &\text{subject to } \left\| z^{\mu} - \boldsymbol{W}^{\kappa}\boldsymbol{\alpha} \right\|_2 \leq w \ \ (k = 1,2,\ldots,2^N; \kappa = 1,2,\ldots,2^{N \cdot M_{\max}}), \\ &\qquad \left\| \boldsymbol{L}\boldsymbol{\alpha} \right\|_2 \leq \sqrt{K}. \end{aligned} \tag{3.25}$$

For our *RMARS* model, we ignore the second constraint of RCQP in Eqs. (3.22), (3.23) and (3.25). Afterwards, we can solve our RCQP by using MOSEK^{TM} [87] software program. Here, we recall that the values $\sqrt{K}$ are determined by a model-free method (cf. Remark 1). When we employ the K values in our RCMARS code and solve by using MOSEK, we apply the K value that has the minimum value of PRSS in Eq. (3.17).

3.1.5 *Numerical Experience with RMARS in the Financial Economics*

As a numerical experiment that may serve to illustrate the implementation of RMARS algorithm developed, in the study [101], we use a small dataset as a sample from the real-world financial market data. It is chosen for our empirical part as time-series data from the website of Central Bank of the Republic of Turkey [25]. The data contain four economic indicators (independent variables) which are the most commonly used ones for the interpretation of an economic situation. These are:

x_1 : ISE Trading Volume, x_2 : Capacity Usage Ratio,

x_3 : Credit Volum, x_4 : Federal Funds Interest Rate.

Here, ISE Trading Volume stands for the number of shares or contracts of a security traded within of a predefined time-window for a month; Capacity Usage Ratio means the ratio of the production capacity of the regarded economy to the maximum capacity of that economy. ISE 100 stock index is the dependent (output or response) variable Y that we try to assess based on our dataset. It consists of 100 stocks that have been chosen among the stocks of companies which are listed on the National Market, and the stocks of real estate investment trusts and venture capital investment trusts, which are listed on the Corporate Products Market. It covers ISE 30 and ISE 50 stocks. As it is a statistical measure of change in an economy or a securities market, we will use that index. For financial markets, an *index* is an imaginary portfolio of securities, representing some market or a portion of it. It possesses its own traditional methods of calculation and, in general, it is represented by a deviation from a base value. Thus, the relative change (in percentage terms) is more important than the absolute value (in actual numerical terms). This dataset

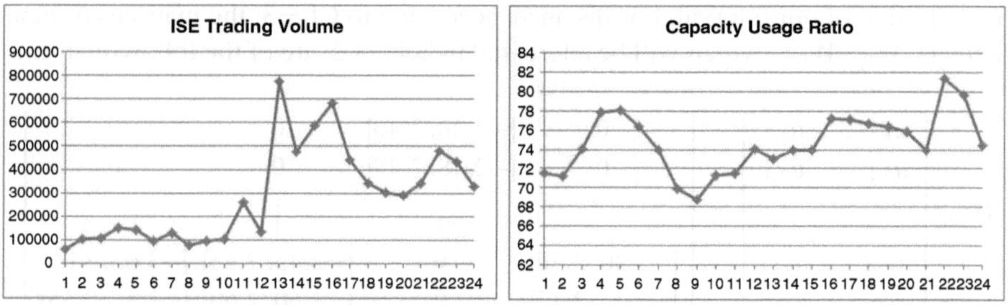

Fig. 3.2 Graph for the characteristic of variables x_1 and x_2

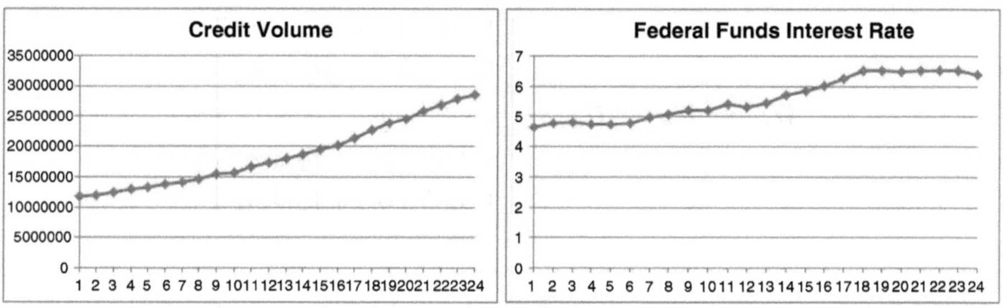

Fig. 3.3 Graph for the characteristic of variables x_3 and x_4

includes 24 observations and the characteristics of our independent variables in time can be seen in Figs. 3.2, 3.3 for a visualization of the dataset.

With this data the largest model is constructed by the forward stepwise stage of Salford MARS Version 3 [83]. After backward stepwise elimination of MARS, the final model is represented as follows:

$$\hat{y} = \alpha_0 + \sum_{m=1}^{M} \alpha_m \vartheta_m(\boldsymbol{x}^m) = \alpha_0 + \alpha_1 \max\{0,\ x_3 - 0.1672\}$$
$$+ \alpha_2 \max\{0,\ 0.1672 - x_3\} + \alpha_3 \max\{0,\ x_4 + 0.4200\}$$
$$+ \alpha_4 \max\{0,\ x_1 + 0.6803\}.$$

To apply the RO technique on MARS model, firstly uncertainties are evaluated for all input values using Eq. (3.16) and all output values. Here these input and output values presented by CIs. Next, we include perturbations (uncertainties) into the real input data $\boldsymbol{x}_k$ in each dimension and into the output data y_k $(k = 1, 2, \ldots, 24)$. For this aim, using Eqs. (3.20)–(3.21), the uncertainty matrices and vectors based on *polyhedral uncertainty sets* are built. Consequently, we construct different uncertainty matrices, $\boldsymbol{U}$ and $\boldsymbol{W}$, for the input data and different uncertainty vectors, $\boldsymbol{z}$ and $\boldsymbol{v}$, for the output data by using six different uncertainty scenarios which are given by the CIs ± 3, $\pm 3/2$, $\pm 3/4$, $\pm 3/6$, $\pm 3/8$ and as a special case, the mid-point (zero) value of our interval. For instance, according to CI ± 3, the matrices of input data, $\boldsymbol{U}$, $\boldsymbol{W}_{up}$, $\boldsymbol{W}_{low}$, which will be addressed in Sect. 3.2, are of the following forms:

$$\boldsymbol{U} = \begin{bmatrix} u_{1,1} & \dots & u_{1,4} \\ u_{2,1} & \dots & u_{2,4} \\ \vdots & \vdots & \vdots \\ u_{23,1} & \dots & u_{23,4} \\ u_{24,1} & \dots & u_{24,4} \end{bmatrix} \in \begin{bmatrix} 0 & [-2.46, 2.46] & 0 & 0 \\ 0 & [-2.48, 2.48] & 0 & 0 \\ \vdots & \vdots & \vdots & \vdots \\ [-2.30, 2.30] & 0 & [-2.45, 2.45] & [-3.05, 3.05] \\ [-2.17, 2.17] & 0 & [-2.60, 2.60] & [-2.55, 2.55] \end{bmatrix},$$

$$\boldsymbol{W}_{up} = \boldsymbol{\vartheta}(\boldsymbol{x}) + \boldsymbol{U}_{up} = \begin{bmatrix} 1 & 0 & 3.89 & 0 & 0 \\ 1 & 0 & 3.89 & 0 & 0 \\ \vdots & \vdots & \vdots & \vdots & \vdots \\ 1 & 3.85 & 0 & 4.12 & 4.42 \\ 1 & 3.85 & 0 & 4.12 & 4.42 \end{bmatrix},$$

$$\boldsymbol{W}_{low} = \boldsymbol{\vartheta}(\boldsymbol{x}) + \boldsymbol{U}_{low} = \begin{bmatrix} 1 & 0 & -1.03 & 0 & 0 \\ 1 & 0 & -1.07 & 0 & 0 \\ \vdots & \vdots & \vdots & \vdots & \vdots \\ 1 & -0.75 & 0 & -0.77 & -1.68 \\ 1 & -0.50 & 0 & -1.07 & -2.68 \end{bmatrix}.$$

Likewise, based on CI ± 3, the uncertainty vectors of output data, $\boldsymbol{z}$, $\boldsymbol{v}_{up}$, $\boldsymbol{v}_{low}$, are represented as follows:

$$\boldsymbol{z} = \begin{bmatrix} z_1 \\ z_2 \\ \vdots \\ z_{23} \\ z_{24} \end{bmatrix} \in \begin{bmatrix} [-3, 3] \\ [-3, 3] \\ \vdots \\ [-3, 3] \\ [-3, 3] \end{bmatrix}, \boldsymbol{v}_{up} = \boldsymbol{y} + \boldsymbol{z}_{up} = \begin{bmatrix} 1.61 \\ 1.76 \\ \vdots \\ 3.52 \\ 2.88 \end{bmatrix}, \boldsymbol{v}_{low} = \boldsymbol{y} + \boldsymbol{z}_{low} = \begin{bmatrix} -4.39 \\ -4.27 \\ \vdots \\ -2.49 \\ -3.12 \end{bmatrix}.$$

As the uncertainty matrix for input data has a very big dimension; and our computer capacity is not enough to solve our problem for this uncertainty matrix (cf. Sect. 3.1.3), we formulate RMARS for each observation using a certain *combinatorial approach* which is called as ***weak robustification***. Therefore, we obtain different *weak RMARS* (*WRMARS*) models to handle that difficulty of complexity. Actually, we have a *tradeoff* between tractability and robustification. As a result, we obtain 24 different WRMARS models and solve them with MOSEK program [87]. Then, we estimate the parameters' values $\alpha_0, \alpha_1, \alpha_2, \alpha_3$ and α_4

Table 3.1 Parameter values and estimation errors of MARS and RMARS

U, v	± 3	$\pm 3/2$	$\pm 3/4$	$\pm 3/6$	$\pm 3/8$	Zero	MARS
α_0	−0.6197	−0.7644	−0.6660	−0.5111	−0.4418	−0.3470	−0.3470
α_1	0.3348	0.2501	−0.3292	−0.5921	−0.7074	−0.8843	−0.8843
α_2	0.0000	0.0000	−0.1852	−0.3686	−0.4494	−0.5722	−0.5722
α_3	0.0000	0.0000	0.4600	0.6262	0.6986	0.8120	0.8120
α_4	0.6529	0.8691	0.7403	0.6508	0.6121	0.5498	0.5498
AAE	0.4048	0.3215	0.1937	0.1385	0.1254	0.1123	0.1123
RMSE	0.4880	0.4204	0.2496	0.1781	0.1559	0.1414	0.1414

Table 3.2 Estimation variance of MARS and RMARS

U, v	± 3	$\pm 3/2$	$\pm 3/4$	$\pm 3/6$	$\pm 3/8$	Zero	MARS
EV	0.447	0.706	0.811	0.88	0.918	0.979	0.979

using a selected WRMARS model which has the *highest* w value in Eq. (3.25) by applying the *worst-case* approach. Finally, we evaluate the regression coefficients and estimation errors based on *Average Absolute Error* (*AAE*) and *Root Mean Squared Error* (*RMSE*) for different uncertainty scenarios. All of the parameter values and estimation errors for MARS and RMARS are represented in Table 3.1.

As we can see in Table 3.1, RMARS produces less accurate results than MARS in terms of AAE and RMSE when the CIs on the variable are very wide. However, as the CIs are narrower, the performance results approach to that of MARS. According to our *main purpose*, we also calculate ***estimation variances*** (***EVs***) for different uncertainty scenarios. EV is the variance of the estimated response values and smaller value of EV provide us the better result. It is evaluated using the following formula:

$$EV := \frac{\sum_{k=1}^{N} (\hat{y}_k - \bar{\hat{y}})}{N-1},$$

where N is number of observations, $\hat{y}_k$ being the kth estimated response value, and $\bar{\hat{y}}$ being the mean of the estimated response values. Based on six different uncertainty scenarios, the values of EV evaluated for our numerical experiment are presented in Table 3.2.

As we may deduce from the results in Table 3.2, RMARS has a much smaller variance than MARS if the CIs on the variable are very wide. As the CIs are narrower, EV increases but, RMARS still has a smaller variance than MARS. Therefore, we can say that RMARS has a considerably smaller EV than MARS for different uncertainty scenarios, as we expect.

While developing RMARS models, a sensitivity study is conducted to define the most suitable confidence limits on both input and output data. For this purpose, different uncertainty matrices for the input data, $\breve{\boldsymbol{x}}$, and different uncertainty vectors for the output data, $\breve{\boldsymbol{y}}$, are obtained by using six different intervals. Above results

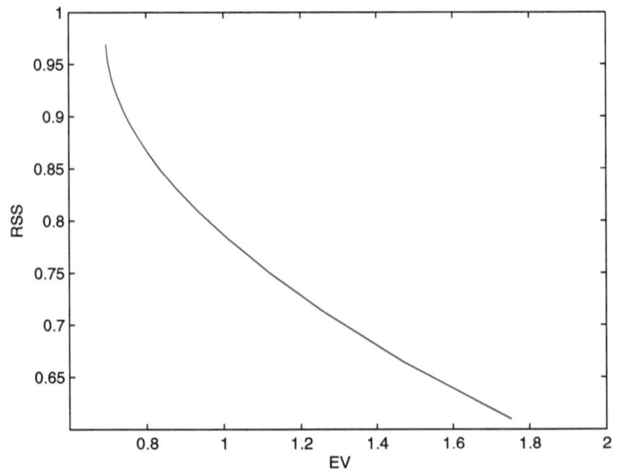

Fig. 3.4 The efficiency frontier between estimation error and estimation variance

in Tables 3.1 and 3.2 indicate that solutions obtained are sensitive to the limits of CIs. When we use the mid-point of our interval values for both input and output data, which is the certain data case (zero interval), we receive the same parameter estimates, and thus, the same model performances and EV values as the ones by MARS. This can disclose that MARS is a ***special case*** of RMARS.

We have a smaller EV when the lengths of the CIs are wide whereas we receive better performance results when the lengths of the CIs are narrow. According to these result, we can observe the tradeoff between *accuracy* (expressed by AAE and RMSE) and *robustness* (given by EV). Also, to analyze this tradeoff clearly, we evaluated the values of Residual Sum of Squares (RSS) and EV based on various different uncertainty scenarios, and we represented the results graphically in Fig. 3.4.

This figure demonstrates the tradeoff between accuracy (given as RSS) and robustness (represented by EV). In economics and finance, this is the so-called *efficiency frontier*.

3.1.6 Simulation Study for RMARS

In the study [101], we compare MARS and RMARS methods using different datasets created by Monte-Carlo simulation based on *variation* of the parameter estimates. Furthermore, in order to see the variation of model performance with parameter estimates, the estimation errors of simulation models are evaluated based on AAE and RMSE. *Monte-Carlo simulation* permits to model situations which present uncertainty and to conduct them many times on a computer. It also includes a risk study through a preparation of models of possible results by substituting a range of values—we may say: a probability distribution—for any factor which implies uncertainty. Monte-Carlo simulation generates distributions of possible outcomes. Through the use of probability distributions, variables may have different probabilities of the outcomes that occur. *Probability distributions* mean a much more realistic kind of representing uncertainty in variables of a risk analysis that belongs to each decision which we are making. Continuously, we are confronted with uncertainty, with ambiguity and variability. It is impossible to precisely estimate the future, even if nowadays we can access an unprecedented amount of information. Monte-Carlo simulation permits us a survey of nearly all the outcomes of our possible decisions and an impact assessment of risk; this allows for a more qualified decision making in the presence of uncertainty [116]. We select normal distribution as a probability distribution to obtain random input variables. For the simulation study of MARS, firstly, we develop a mathematical model. This model is the *process model* and represented as follows:

$$Y = -3050 + 0.02x_1 + 50x_2 - 0.0009x_3 + 8400x_4 + 30x_2x_4 + \varepsilon. \tag{3.26}$$

Afterward, using Minitab package program [84] generated random input variables chosen from suitable distribution function which are expected to determine the variables. Here we simulate values of a *normal random variable*. Then, using Eq. (3.26), we monitor preferred output variables which become distributions whose properties are described by the model and the distributions of the random variables. So, we generate 30 different simulated datasets to employ simulation for MARS and 30 different MARS models are constructed using Salford Systems MARS [83]. In fact, the parameter values of MARS models are estimated according to these simulated datasets. Some selected MARS models obtained are of the following form:

$$\hat{y} = \alpha_0 + \alpha_1 \max\{0,\ x_1 - 2.25\} + \alpha_2 \max\{0,\ x_3 + 1.86\} + \alpha_3 \max\{0,\ x_4 + 2.18\},$$

$$\hat{y} = \alpha_0 + \alpha_1 \max\{0,\ x_1 - 1.47\} + \alpha_2 \max\{0,\ x_3 + 2.84\} + \alpha_3 \max\{0,\ x_4 + 2.45\},$$

$$\hat{y} = \alpha_0 + \alpha_1 \max\{0,\ x_1 - 3.07\} + \alpha_2 \max\{0,\ x_3 + 1.78\} + \alpha_3 \max\{0,\ x_4 + 2.63\},$$

$$\hat{y} = \alpha_0 + \alpha_1 \max\{0,\ x_1 - 2.39\} + \alpha_2 \max\{0,\ x_3 + 1.62\} + \alpha_3 \max\{0,\ x_4 + 1.51\},$$

$$\hat{y} = \alpha_0 + \alpha_1 \max\{0,\ x_1 - 2.21\} + \alpha_2 \max\{0,\ x_3 + 1.98\} + \alpha_3 \max\{0,\ x_4 + 3.27\},$$

$$\hat{y} = \alpha_0 + \alpha_1 \max\{0,\ x_1 - 1.98\} + \alpha_2 \max\{0,\ x_3 + 2.07\} + \alpha_3 \max\{0,\ x_4 + 1.49\},$$

$$\hat{y} = \alpha_0 + \alpha_1 \max\{0,\ x_1 - 2.74\} + \alpha_2 \max\{0,\ x_3 + 2.20\} + \alpha_3 \max\{0,\ x_4 + 1.70\}.$$

For simulation study of RMARS, firstly, 30 different interval values are determined and, hence, under polyhedral uncertainty sets, thirty different uncertainty scenarios are obtained by using these values. The values of the CIs are $\pm 3/2$, $\pm 3/2.1$, $\pm 3/2.2, \ldots$, $\pm 3/4.6$, $\pm 3/4.8$, $\pm 3/5$. Then the RMARS model frames are constructed by running a MATLAB code written by us and MOSEK software [87] is

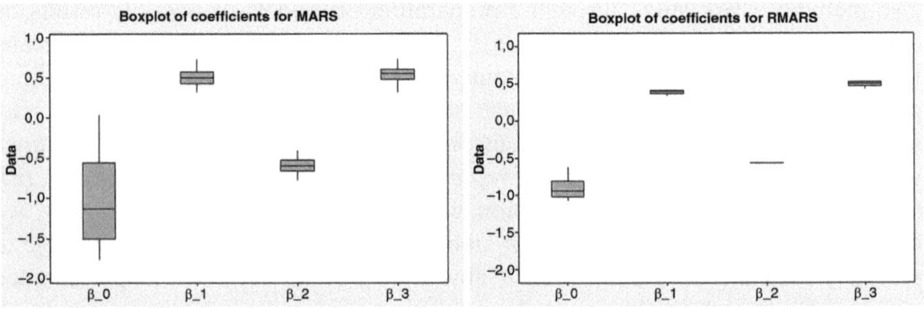

Fig. 3.5 The graphical representation for the variance of parameter estimates of MARS and RMARS

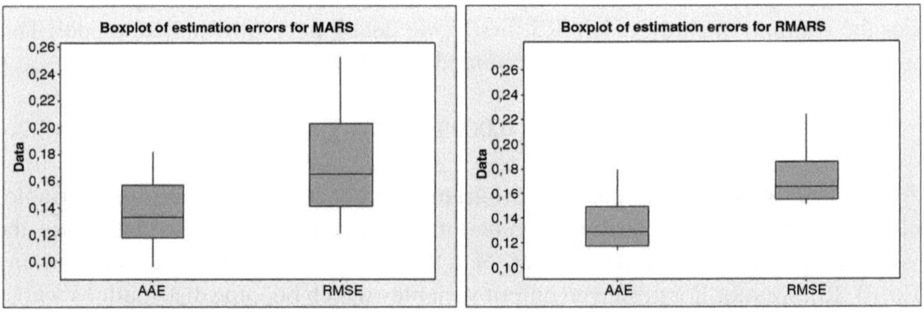

Fig. 3.6 Based on AAE and RMSE, the graphical representation for the variance of model performance criteria of MARS and RMARS

used to solve the CQP problems for RMARS models. Hereby, parameter values of RMARS models are also predicted based on 30 different uncertainty scenarios.

In RMARS, our aim is to decrease the *estimation variance* by implying full robustification in MARS even though the estimation errors of RMARS can be slightly higher than those of MARS when we incorporate perturbation (uncertainty) using Eq. (3.16) into the input and output data based on polyhedral uncertainty sets, defined in Eqs. (3.20)–(3.21). For this simulation study, the results for the variance of parameter estimates can be understood by Fig. 3.5.

As we observe in Fig. 3.5, the variability of the model parameter estimates of the RMARS is much less than that of MARS. For this simulation study, the variance of model performance can be learned from Fig. 3.6 for our two methods.

As we can deduce from Fig. 3.6, similar to the variability of the model parameter estimates, the variability of model performance in terms of estimation errors of the RMARS is less than that of MARS.

Remark 3 In Figs. 3.5, 3.6, we just give a graphical representation based on MARS and RMARS with considering that we receive similar results for CMARS and RCMARS.

3.2 Robust (Conic) Generalized Partial Linear Models

3.2.1 Introduction

In previous sections, we upgraded the (C)MARS model to be able to model the datasets including uncertainty for future scenarios and constructed R(C)MARS method with the help of robust optimization. Although, in the real life, variables are generally nonlinearly implied and, in this case, we need nonlinear models that can minimize the error term, knot selection is an extra important point in terms of complexity to solve the problem in (C)MARS and R(C)MARS and it is not necessary for their linear part. Since a dataset can contain linear and nonlinear variables and linear regression is very successful in determining a linear relationship between the variables, in this section, a new model, R(C)GPLM is presented with essential contributions of R(C)MARS and Linear (or Logistic) Regression as a developed version of GPLM to partially reduce the complexity of R(C)MARS.

GPLM [90] is a combination of two different regression models each of which is used to apply on different parts of the data set. Generalized Linear Models (GLMs) have been advanced to constitute GPLMs enlarging the linear terms through a nonlinear component, '*P*' meaning partial. Such semiparametric models are needed to develop, because of the inflexibility of simple linear and nonlinear models to show the trends, relations and anomalies buried in real-life datasets. GPLM is adequate to high dimensional, non-normal and nonlinear data sets having the flexibility to reflect all anomalies effectively. In the study [137], Conic GPLM (CGPLM) was introduced using CMARS and Logistic Regression. According to a comparison with CMARS, CGPLM gave better results. In the studies [99, 100], we include the existence of uncertainty in the future scenarios into (C)MARS and the linear/logit regression parts in (C)GPLM and we robustify entire terms with robust optimization which is dealt with data uncertainty.

3.2.2 General Description of (C)GPLM

GPLMs apply a bipartite model separately on linear and nonlinear parts, and they have a general form [90]:

$$E\left(Y|\boldsymbol{X},\boldsymbol{T}\right)=G\left(\boldsymbol{X}^{T}\boldsymbol{\beta}+\varsigma\left(\boldsymbol{T}\right)\right), \tag{3.27}$$

where the vectors $\boldsymbol{X}$ and $\boldsymbol{T}$ represent a decomposition of variables whose parameters and further unknowns would be assessed by linear and nonlinear models, respectively. Furthermore, $\varsigma(\cdot)$ is a smooth function estimated for the nonlinear model. Finally, a link function $G = H^{(-1)}$ makes the connection between the mean of the dependent variable $\mu = E(Y|\boldsymbol{X}, \boldsymbol{T})$ and the predictor variables:

$$H(\mu) = \nu\left(\boldsymbol{X}, \boldsymbol{T}\right) = \boldsymbol{X}^T\boldsymbol{\beta} + \varsigma\left(\boldsymbol{T}\right) = \sum_{j=1}^{m} X_j\beta_j + \varsigma\left(\boldsymbol{T}\right). \tag{3.28}$$

In (C)GPLM, the LR model constitutes the linear part of the estimation:

$$Y = \beta_0 + \sum_{j=1}^{p} X_j\beta_j + \varepsilon, \tag{3.29}$$

where β_0 is the intercept term, β_j are the coefficients of X_j $(j = 1, 2, \ldots, p)$, ε is the noise term, Y is the dependent variable and X_j are the independent variables.

In the nonlinear part of the (C)GPLM [30, 31, 63, 137], a new variable $Y^{preproc}$ is defined by the help of β_0, β_j and the variables X_j $(j = 1, 2, \ldots, p)$, which would determine the knots of MARS on the residuals, with q nonlinearly involved variables T_j that are not used in the linear part:

$$Y^{preproc} = \boldsymbol{X}\boldsymbol{\beta}^{preproc} + \varepsilon = \beta_0 + \sum_{j=1}^{p} X_j\beta_j + \varepsilon. \tag{3.30}$$

After the evaluation of linear part and getting the regression coefficients' vector $\boldsymbol{\beta}^{preproc}$, which is an optimal vector found as a result of the linear least-squares model, the residual $\hat{\boldsymbol{y}}$ is defined by the subtraction of $\boldsymbol{X}\boldsymbol{\beta}^{preproc}$ from $\boldsymbol{y}$:

$$\boldsymbol{y} - \boldsymbol{X}\boldsymbol{\beta}^{preproc} =: \boldsymbol{\gamma}. \tag{3.31}$$

In Eq. (3.31), $\boldsymbol{y}$ is the given response data vector, $\boldsymbol{\gamma}$ is the resulting vector of residuals which is constructed to develop the knot selection by MARS and apply the backward process with CMARS. Furthermore, in Eq. (3.31), $\boldsymbol{X}$ stands for the design matrix of input data due to the linear model.

The smooth function $\varsigma(\cdot)$ of GPLM is estimated by (C)MARS during the forward process. This function, which is a linear combination of basis functions ϑ_m and the intercept α_0, can be represented by MARS and also the alternative model of CMARS that is considered as a substitute of the backward process of MARS:

$$\boldsymbol{\gamma} = H(\mu) = \sum_{m=1}^{M} \alpha_m\vartheta_m(\boldsymbol{t}^m). \tag{3.32}$$

We note that the "bias" term α_0 is not needed in Eq. (3.32) since it already entered as a part, namely, β_0, of the linear model. Next, for the alternative model of CMARS as a component of the nonlinear part of CGPLM, the Penalized Residual Sum of

Square (PRSS) in Eq. (3.6) will be constructed by the equation:

$$PRSS := \sum_{k=1}^{N}(\gamma_k - \boldsymbol{\alpha}^T \boldsymbol{\vartheta}(\boldsymbol{b}_k))^2 + \sum_{m=1}^{M_{\max}} \phi_m \sum_{\substack{|\theta|=1 \\ \theta^T=(\theta_1,\theta_2)}}^{2} \sum_{\substack{r<s \\ r,s\in V(m)}} \int_{Q^m} \alpha_m^2 [D_{r,s}^{\theta} \vartheta_m(\boldsymbol{t}^m)]^2 d\boldsymbol{t}^m. \tag{3.33}$$

The multi-dimensional integrals of PRSS are approximated by using the discretization to represent it as follows, where $\phi \geq 0$:

$$PRSS \approx \left\| \boldsymbol{\gamma} - \boldsymbol{\vartheta}(\boldsymbol{b})\boldsymbol{\alpha} \right\|_2^2 + \phi \left\| \boldsymbol{L}\boldsymbol{\alpha} \right\|_2^2. \tag{3.34}$$

3.2.3 Robustification of (C)GPLM

Here, we include the existence of uncertainty in the future scenarios into (C)GPLM, which can be represented in the following form [99, 100]:

$$E(Y|\boldsymbol{X},\boldsymbol{T}) = G(\underbrace{\boldsymbol{X}^T}_{\text{noisy variable}} \boldsymbol{\beta} + \varsigma(\underbrace{\boldsymbol{T}}_{\text{noisy variable}})), \tag{3.35}$$

where $\boldsymbol{X} = (X_1, X_2, \ldots, X_p)^T$ and $\boldsymbol{T} = (T_1, T_2, \ldots, T_q)^T$ are a decomposition of variables, when $\boldsymbol{X}$ denotes an p-variate vector showing the variables with a linear pattern, $\boldsymbol{T}$ denotes a q-variate vector showing the variables with a nonlinear pattern, to be estimated with a nonlinear model. In Eq. (3.35), $\boldsymbol{\beta} = (\beta_1, \beta_2, \ldots, \beta_p)^T$ consists of the coefficient vector of $\boldsymbol{X}$, estimated by a linear (or logit) regression model, and $\varsigma(\cdot)$ is a smooth function estimated by a nonlinear model. In this study, we focus on special types of estimation $\varsigma(\cdot)$ by R(C)MARS.

The variables X_j $(j = 1, 2, \ldots, p)$ and T_j $(j = 1, 2, \ldots, q)$ are supposed to be normally distributed random variables. For each input variable X_j and T_j, a transformation is made through uncertainties:

$$\begin{aligned} \breve{X}_j &= \bar{x}_j + \xi_j \quad (j = 1, 2, \ldots, p), \\ \breve{T}_j &= \bar{t}_j + \zeta_j \quad (j = 1, 2, \ldots, q). \end{aligned} \tag{3.36}$$

To robustify (C)GPLM, with similar idea of R(C)MARS, we apply robust optimization on linear and nonlinear parts in the (C)GPLM, and, in Eq. (3.36), we assume that the input and output variables of our (C)GPLM are represented by random variables. They lead us to uncertainty sets, which are assumed to contain confidence intervals (CIs) [101]. We incorporate a 'perturbation' (uncertainty) into the real input data $(\boldsymbol{x}_k, \boldsymbol{t}_k)$ in each dimension, and into the output data y_k. Therefore, our new values of R(C)GPLM are shown in the following:

$$\begin{aligned}
&x_{k,j} \to \breve{x}_{k,j}; \quad \breve{x}_{k,j} = \bar{x}_j + \delta_{k,j}, \quad |\delta_{kj}| \le \upsilon_{kj} \quad (k = 1,2,\ldots,N; j = 1,2,\ldots,p),\\
&t_{k,j} \to \breve{t}_{k,j}; \quad \breve{t}_{k,j} = \bar{x}_j + \Delta_{k,j}, \quad |\Delta_{kj}| \le \rho_{kj} \quad (k = 1,2,\ldots,N; j = 1,2,\ldots,q),\\
&y_k \to \breve{y}_k; \quad \breve{y}_k = \bar{y}_k + \tau_k, \quad |\tau_k| \le \nu_k \quad (k = 1,2,\ldots,N).
\end{aligned}$$

With the uncertainty sets $U_1^1 \subseteq \mathbb{R}^{N \cdot p}, U_1^2 \subseteq \mathbb{R}^{N \cdot M_{max}}$ and $U_2^1, U_2^2 \subseteq \mathbb{R}^{N \cdot}$ applied on the data $(\breve{\boldsymbol{x}}_k, \breve{y}_k)$ and $(\breve{\boldsymbol{t}}_k, \breve{y}_k)$ $(k = 1,2,\ldots,N)$, our model of Eq. (3.28) implies uncertainty, can be represented as an additive semiparametric model:

$$H(\breve{\mu}) = \nu\left(\breve{\boldsymbol{X}}, \breve{\boldsymbol{T}}\right) = \breve{\boldsymbol{X}}^T \boldsymbol{\beta} + \varsigma\left(\breve{\boldsymbol{T}}\right) = \sum_{j=1}^{p} \breve{X}_j \beta_j + \varsigma\left(\breve{\boldsymbol{T}}\right). \tag{3.37}$$

The observation value and vectors, $\breve{y}_k, \breve{\boldsymbol{x}}_k, \breve{\boldsymbol{t}}_k$ $(k = 1,2,\ldots,N)$ with uncertainty, respectively, $\breve{\mu} = G(\breve{\boldsymbol{\gamma}}_k)$ and $\breve{\gamma}_k = H(\breve{\mu}_k) = \breve{\boldsymbol{x}}_k^T \boldsymbol{\beta} + \varsigma(\boldsymbol{t}_k)$ with a smooth function $\varsigma(\cdot)$ are considered in the form of a RCGPLM.

3.2.4 Linear (Logit) Regression Model for the Linear Part

In the linear part of the estimation, a new variable $Y^{preproc}$ is constructed by the help of the coefficients β_0, β_j and $\breve{X}_j$ $(j = 1,2,\ldots,p)$. This variable would be later used in MARS with reminded nonlinear variables $\breve{T}_j$ $(j = 1,2,\ldots,q)$ on residuals to determine the knot values [99, 102]:

$$Y^{preproc} = \breve{\boldsymbol{X}}^T \boldsymbol{\beta}^{preproc} + \varepsilon = \beta_0 + \sum_{j=1}^{p} \breve{X}_j \beta_j + \varepsilon. \tag{3.38}$$

With an appropriate bound of K, LR model may be solved with a continuous optimization technique, CQP and have the following form:

$$\begin{aligned}
&\underset{w_1, \boldsymbol{\beta}}{\text{minimize}} \quad w_1\\
&\text{subject to} \quad \left\|\boldsymbol{y} - \breve{\boldsymbol{X}}\boldsymbol{\beta}\right\|_2 \le w_1,\\
&\qquad\qquad\quad \left\|\boldsymbol{L}_1 \boldsymbol{\alpha}\right\|_2 \le \sqrt{K_1}.
\end{aligned} \tag{3.39}$$

For our RGPLM model, we ignore the second constraint of RCQP in Eq. (3.39). To obtain the response variable $\breve{\boldsymbol{\gamma}}$ for the nonlinear part, the same procedure with Sect. 3.2.2 can be applied.

3.2.5 R(C)MARS Method for the Nonlinear Part

The smooth function $\varsigma(\cdot)$ to be estimated by R(C)MARS in the Eq. (3.28), is represented as a linear combination of basis functions ϑ_m to transform the model of Eq. (3.31) into the form:

$$\gamma = H(\mu) = \sum_{m=1}^{M} \alpha_m \vartheta_m(t^m). \tag{3.40}$$

Then, for RCGPLM, PRSS in Eq. (3.34) with uncertainty can be converted into the following form:

$$PRSS \approx \left\| \breve{y} - \boldsymbol{\vartheta}(\breve{t})\boldsymbol{\alpha} \right\|_2^2 + \phi \left\| \boldsymbol{L\alpha} \right\|_2^2. \tag{3.41}$$

With an appropriate bound of K_1, PRSS can be easily solved with our continuous optimization technique, CQP, in terms of TR, and have the subsequent form [99]:

$$\begin{aligned} &\underset{w_2,\boldsymbol{\alpha}}{\text{minimize}} \quad w_2 \\ &\text{subject to} \quad \left\| \boldsymbol{y} - \boldsymbol{\vartheta}(\boldsymbol{b})\boldsymbol{\alpha} \right\|_2 \leq w_2, \\ &\qquad\qquad\quad \left\| \boldsymbol{L\alpha} \right\|_2 \leq \sqrt{K_2}. \end{aligned} \tag{3.42}$$

We underline that we receive robust CQP for RGPLM model, we do not consider the second constraint in Eq. (3.42).

3.2.6 R(C)GPLM with Polyhedral Uncertainty

In this subsection, the form of polyhedral as uncertainty sets is employed to be able to continue our study with standard CQP.

3.2.6.1 Robust Counterpart for Linear Part

In this part, uncertainty is constructed by a family of matrices $\breve{\boldsymbol{W}} = W + \boldsymbol{U}_1$ and vectors $\breve{z} = \boldsymbol{z} + \boldsymbol{v}_1$, where $\boldsymbol{U}_1 \in U_1^1$ and $\boldsymbol{v}_1 \in U_2^1$ are unknown and lying in bounded uncertainty sets represented in Sect. 3.1.3, with the semilengths ρ, ν_1 of confidence sets, respectively. As we use polyhedral uncertainty for the linear part of RCGPLM, with the uncertainty sets U_1^1 and U_2^1, the robust counterpart can be expressed by [102]

$$\underset{\boldsymbol{\beta}}{\text{minimize}} \max_{\substack{W \in U_1^1, \\ z \in U_2^1}} \left\| \boldsymbol{\beta} - \boldsymbol{W} z \right\|_2^2 + \phi \left\| \boldsymbol{L}_1 \boldsymbol{\beta} \right\|_2^2, \tag{3.43}$$

with the polytopes U_1^1 and U_2^1 described by their vertices:

$$U_1^1 = \text{conv}\{\boldsymbol{W}_1^1, \boldsymbol{W}_1^2, \ldots, \boldsymbol{W}_1^{2^{N \cdot p}}\}, \; U_2^1 = \text{conv}\{z_1^1, z_1^2, \ldots, z_1^{2^N}\}. \tag{3.44}$$

The linear part of RCGPLM can be represented as a standard CQP problem [12, 102]:

$$\begin{aligned} &\underset{w_1, \boldsymbol{\beta}}{\text{minimize}} && w_1 \\ &\text{subject to} && \left\| z_1^{\mu} - \boldsymbol{W}_1^{\kappa} \boldsymbol{\beta} \right\|_2 \leq w_1 \; (\mu = 1, 2, \ldots, 2^N; \kappa = 1, 2, \ldots, 2^{N \cdot p}), \\ &&& \left\| \boldsymbol{L}_1 \boldsymbol{\alpha} \right\|_2 \leq \sqrt{K_1}, \end{aligned} \tag{3.45}$$

where $K_1 \geq 0$ is an appropriate bound value. We recall that, along of the parameter K_1, we obtain an efficiency frontier of solutions of Eq. (3.45), where a special selection can be chosen via statistical and, further performance and comparison criteria.

For linear part in the *RGPLM* model, we just have to drop the second part in Eqs. (3.43) and (3.45).

3.2.6.2 Robust Counterpart for Nonlinear Part

Here, uncertainty is constructed by a family of matrices $\vartheta(\breve{\boldsymbol{t}}) = \vartheta(\boldsymbol{t}) + \boldsymbol{U}_2$ and vectors $\breve{\boldsymbol{\gamma}} = \boldsymbol{\gamma} + \boldsymbol{v}_2$, where $\boldsymbol{U}_2 \in U_1^2$ and $\boldsymbol{v}_2 \in U_2^2$ within bounded uncertainty sets, identified in Sect. 3.1.3, with the semilengths ρ, ν_2 of our confidence sets, respectively.

When we use polyhedral uncertainty for the nonlinear part of CGPLM, with the uncertainty sets U_1^2 and U_2^2, the robust counterpart can be represented as

$$\underset{\boldsymbol{\alpha}}{\text{minimize}} \max_{\substack{W \in U_1^2, \\ z \in U_2^2}} \left\| z_2 - \boldsymbol{W}_2 \boldsymbol{\alpha} \right\|_2^2 + \phi \left\| \boldsymbol{L}_2 \boldsymbol{\alpha} \right\|_2^2, \tag{3.46}$$

with the polytopes

$$U_1^1 = \text{conv}\{\boldsymbol{W}_2^1, \boldsymbol{W}_2^2, \ldots, \boldsymbol{W}_2^{2^{N \cdot M_{\max}}}\}, U_2^1 = \text{conv}\{z_2^1, z_2^2, \ldots, z_2^{2^N}\}. \tag{3.47}$$

We can express our robust problem of Eq. (3.46) as a standard CQP problem:

$$
\begin{aligned}
&\underset{w_2,\boldsymbol{\alpha}}{\text{minimize}} && w_2 \\
&\text{subject to} && \left\| \boldsymbol{z}_{\mathbf{1}}^{\mu} - \boldsymbol{W}_{\mathbf{1}}^{\kappa}\boldsymbol{\alpha} \right\|_2 \leq w_2 \ (\mu = 1,2,\ldots,2^N; \kappa = 1,2,\ldots,2^{N \cdot M_{\max}}), \\
& && \left\| \boldsymbol{L}_2\boldsymbol{\alpha} \right\|_2 \leq \sqrt{K_2},
\end{aligned}
\tag{3.48}
$$

where $K_2 \geq 0$ is an appropriate bound value. We recall that, for the nonlinear part of our *RGPLM* model, we have not taken into consideration the second part in Eqs. (3.46) and (3.48).

Chapter 4
Spline Regression Models for Complex Multi-Model Regulatory Networks

In the previous chapter, we gave some details on theory and methods of regression and classification, (C)MARS, and their robust counterpart, R(C)MARS. and we represented and applied our methods to real-world data from different sectors. In this chapter, we apply the data mining tool of regression and classification, (C)MARS, on a dynamics. By this, the amount of condition grows, since each time point (a discrete time, in our case) can be regarded as an extra 'condition'; in this way, there would be unknown parameters needed in order to balance the number of constraints, i.e., to close the gap of 'degree of freedom'. In this respect, the number of unknown parameters would need to be relatively high, necessarily. However, in our research, we try to gain from the dataset topologically and geometrically best, to 'get into' the dynamics smartly, benefiting from structural features of the dataset. In this respect, the algorithm of MARS and CMARS seems to be an excellent choice as, e.g., in each dimension of the input variables, we get a piecewise linear 'zig-zag' function, where the linear parts present and approximate the data over whole intervals. This process is done adaptively, which also means: smartly.

We note that the use of CMARS instead of MARS allows for an integrated representation of the entire parameter identification task as an optimization program in the sense of a model-based problem rather than a model-free one. By this CMARS permits to employ the rigor of optimization theory and it also gives a chance for future generalizations of this research which might benefit from further areas of optimization theory, such as Stochastic Programming and, especially, RO. In fact, our newly developed RCMARS aims at a rigorous regularization not only with respect to the output variables but also in the input variables, we might say: in the design of the program, with the help of Robust Optimization. For any case of such further extensions which is represented in following chapter, we need not newly

A. Özmen, *Robust Optimization of Spline Models and Complex Regulatory Networks*, Contributions to Management Science,
DOI 10.1007/978-3-319-30800-5_4

return from any model freeness to an optimization theoretical model, since we are here in a model-based setting already.

Since we regularize the model of CMARS, including first- and second-order derivatives, we go for 'easy' models, by penalization; we turn this regularization into the mathematical language of CQP. One expression of that easiness is the—to some sense enforced (entire complexity bounded)—'flatness' of the model. In tendency, we can say: we force the components of the vector of unknown parameters to be as much as possible nearby to vanish. But this also means: we try to have as small as possible a number of 'significant' parameters in our model. MARS does this with the help of an "index" called GCV (cf. Sect. 2.3.3) we, with CMARS, do this by a more integrated optimization theory framework. We remark that these reflections hold true for the two classes of variables, respectively, namely, the target variables and the environmental variables. Furthermore, the introduction of the environment and its items themselves into the model, in addition to the target variables, already means some kind of regularization, i.e., a reduction of complexity. In fact, the environmental variables take away from the target variables some of the huge modeling load to explain 'alone' the data accurately by the model.

Another class of parameters in this chapter are the penalty parameters or, in terms representation, equivalently, upper bounds of the complexity. We already reduced these parameters just by single values, per class of variables, and not per basis functions individually. This means a strong reduction of the entire numbers of parameters. However, we do still have the option of employing these parameters further in a more refined, individual manner, depending on the entire model and its complexity. In this respect, we can 'tune'. In fact, we would like to mention that this work on the number of parameters can also be called as a *model selection*, including the suitable choice of dimensions. Regarding the choice of the upper bound parameters, we have an experience in the use of statistical 'performance' or 'comparison' measures through a number of research works on CMARS and its robust counterpart RCMARS.

Finally, the knot points are another large group of further parameters. In MARS, these knots are selected automatically in a forward stepwise manner when fitting a MARS model. We may also approach them from the perspective of a 'splitting' between the *classes* in each input dimension which reveals a large variation between the classes, as we know it from the famous classification method CART [24]. In CMARS, we propose to choose the knot points projectively, in each dimension, nearby to the data points or, to be more precise, to the grid points canonically generated by the data points.

We underline that all these intentions and efforts to improve aim at an accurate and, at the same time, 'double', not too complex but for future applications well-prepared methodology of CMARS and of its emerging and forthcoming varieties. In this chapter, we represent and investigate a ***'dynamical counterpart'*** of this research agenda. We analyze time-discrete target-environment regulatory systems (TE systems) with spline entries, and we present and solve new regression problems by using MARS and CMARS. We apply these methods on small artificial datasets which have 4 variables (2 targets and 2 environmental factors) and 25 samples, as our numerical experience prepared. We also obtain a simulation study based on 5 different datasets and compare the performances of MARS and CMARS.

4.1 Regression Problem for Regulatory Network with Spline Entries

4.1.1 Introduction

Here, the use of spline function possesses great and, in fact, invaluable advantages, in general, and, especially, in the context or our modeling of a dynamics [103]. Indeed,

(*i*) Splines, from the viewpoint of a single dimension (input variable), are piecewise polynomials. If we just used polynomials, then they would usually converge to plus or minus infinity when the absolute values of the (input) variables grow large. As many real-world processes generally stay in bounded margins even if these bounds and the time horizons are very large, polynomials would need to be of a high degree to 'turn around' or oscillate enough in order to stay in that margin. But with high-degree polynomials it is not that easy to work, especially, since the real-world problems are multivariate, which can imply multiplication effects and, hence, a fast increase of the degree of the occurring multidimensional splines. Instead, using our elementary (C)MARS splines allows us, in each dimension, to keep the degree of the polynomial 'pieces' very low. The splines are quite 'flexible' indeed, such to say, 'elastic'. Often, we call splines 'smoothing splines' as they 'smoothly' approximate the discrete data.

(*ii*) Splines of CMARS are even more 'smooth' as their oscillatory behavior is kept under control through a penalization of their complexity (integral of squared first- and, in particular, second-order derivatives); then we discretize the integral, receive a problem of Tikhonov regularization which we finally represent as a problem of CQP.

(*iii*) The multivariate splines of (C)MARS are products of 'zig-zagging', i.e., piecewise linear functions, which are piecewise of degree 1 (or 0), and we can carefully decide on how many dimensions we include into the process of multiplications of these 1-dimensional splines. In fact, both the low 1-dimensional degrees and the controlled multiplication amounts to an additional care about that the complexity of our model will not be too high. We recall that a reduction of complexity may also be named an increase of stability.

(*iv*) That we perform those multiplications is an expression of the fact that the input variables are dependent and together, in groups, contribute to an explanation of the response variable by those explanatory input variables.

(*v*) Finally, differently from the use of a 'stiff' model formula (which are motivated by the tradition of physical sciences), our approach by CMARS is very adaptive and is getting "into" the dataset with its particular subsets and characteristics of shape.

Therefore, in the study [103], we introduce a regulatory system with (C)MARS spline entries. Our research on regulatory systems started with the assessment of the dynamics of genetic networks, gene-environment networks and eco-finance networks. Those dynamical models were introduced in the time-continuous version first, and then treated time-discretely; careful discussions on the time-continuous vs. -discrete nature of the dynamical model were made. By that we move from MARS to the more 'continuous' (in terms of the model and of the continuous optimization methods used) alternative CMARS, we are staying closer to the originally continuous nature of the subject of our study.

4.1.2 The Dynamical Procedure

Selecting the entries of the matrix that encode our regulatory network as splines, MARS, or alternatively, CMARS can be used to find the unknown parameters in TE networks. By inserting splines in Eq. (2.11), we obtain the following predictions [103]:

$$\begin{aligned} \widehat{X}_j^{(k+1)} &= \alpha_{j0}^{\mathrm{T}} + \vartheta_{\alpha_j^{\mathrm{TT}}}\big(\widetilde{\boldsymbol{X}}^{(k)}\big) + \vartheta_{\alpha_j^{\mathrm{ET}}}\big(\widetilde{\boldsymbol{E}}^{(k)}\big), \\ \widehat{E}_i^{(k+1)} &= \alpha_{i0}^{\mathrm{E}} + \vartheta_{\alpha_i^{\mathrm{TE}}}\big(\widetilde{\boldsymbol{X}}^{(k)}\big) + \vartheta_{\alpha_i^{\mathrm{EE}}}\big(\widetilde{\boldsymbol{E}}^{(k)}\big). \end{aligned} \tag{4.1}$$

When we compare measurements and predictions and use the (Euclidean) $\|\cdot\|_2$-norm, we can identify our model by solving the following least-squares (or in some probabilistic setting, maximum-likelihood) estimation problem:

$$\text{minimize} \sum_{k=0}^{N} \left(\big\|\widehat{\boldsymbol{X}}^{(k)} - \widetilde{\boldsymbol{X}}^{(k)}\big\|_2^2 + \big\|\widehat{\boldsymbol{E}}^{(k)} - \widetilde{\boldsymbol{E}}^{(k)}\big\|_2^2 \right).$$

After using the form of BFs in Eq. (3.4) and adding penalty terms in the regression model of TE networks to control the lack of fit from the viewpoint of the complexity and stability, the discretized form of PRSS in Eq. (3.6) can be approximately represented as follows:

$$PRSS \approx \big\|\widehat{\boldsymbol{X}} - \widetilde{\boldsymbol{X}}\big\|_2^2 + \big\|\widehat{\boldsymbol{E}} - \widetilde{\boldsymbol{E}}\big\|_2^2 + \phi_{\mathrm{T}} \left\|\boldsymbol{L}_{\mathrm{T}}\boldsymbol{\alpha}_{\mathrm{T}}\right\|_2^2 + \phi_{\mathrm{E}} \left\|\boldsymbol{L}_{\mathrm{E}}\boldsymbol{\alpha}_{\mathrm{E}}\right\|_2^2. \tag{4.2}$$

By this representation, the PRSS minimization problem looks like a classical TR problem and it can be coped with CQP [96, 136]. Using suitable bounds K_{T} and K_{E}, we may rewrite our optimization problem in the subsequent form:

$$\begin{aligned} &\underset{w_{\mathrm{T}}, w_{\mathrm{E}}, \boldsymbol{\alpha}}{\text{minimize}} \;\; w_{\mathrm{T}} + w_{\mathrm{E}} \\ &\text{subject to } \left\| \widehat{\boldsymbol{X}} - \widetilde{\boldsymbol{X}} \right\|_2 \leq w_{\mathrm{T}} \quad \text{(cf. Eqn. (4.1))}, \\ &\qquad \left\| \widehat{\boldsymbol{E}} - \widetilde{\boldsymbol{E}} \right\|_2 \leq w_{\mathrm{E}} \quad \text{(cf. Eqn. (4.1))}, \\ &\qquad \|\boldsymbol{L}_{\mathrm{T}} \boldsymbol{\alpha}_{\mathrm{T}}\|_2 \leq \sqrt{K_{\mathrm{T}}}, \\ &\qquad \|\boldsymbol{L}_{\mathrm{E}} \boldsymbol{\alpha}_{\mathrm{E}}\|_2 \leq \sqrt{K_{\mathrm{E}}}. \end{aligned} \tag{4.3}$$

However, to simplify our model in Eq. (4.2) by a single penalty parameter ϕ, PRSS can be approximately given as follows:

$$PRSS \approx \overbrace{\left\| \widehat{\boldsymbol{X}} - \widetilde{\boldsymbol{X}} \right\|_2^2}^{=Accuracy} + \overbrace{\phi \, \|\boldsymbol{L}\boldsymbol{\alpha}\|_2^2}^{\approx Complexity}, \tag{4.4}$$

where $\boldsymbol{\alpha} = (\boldsymbol{\alpha}_{\mathrm{T}}^T, \boldsymbol{\alpha}_{\mathrm{E}}^T)^T$ is an $((M_{\max+1}) \times 1)$-parameter vector, to be estimated through the given data points. Here, $\widetilde{\boldsymbol{X}} = (\widetilde{\boldsymbol{X}}^T, \widetilde{\boldsymbol{E}}^T)^T$, $\widehat{\boldsymbol{X}} = (\widehat{\boldsymbol{X}}^T, \widehat{\boldsymbol{E}}^T)^T$and $\boldsymbol{L} = (\boldsymbol{L}_{\mathrm{T}}, \boldsymbol{L}_{\mathrm{E}})$. Therefore, for target-environment networks, we may present our optimization problem as given below:

$$\begin{aligned} &\underset{w, \boldsymbol{\alpha}}{\text{minimize}} \quad w \\ &\text{subject to} \quad \left\| \widehat{\boldsymbol{X}} - \widetilde{\boldsymbol{X}} \right\|_2 \leq w \quad \text{(cf. Eqn. (4.1))}, \\ &\qquad\qquad \|\boldsymbol{L}\boldsymbol{\alpha}\|_2 \leq \sqrt{K}. \end{aligned} \tag{4.5}$$

In CMARS approach, via the 'control parameter' given by some upper bound of the complexity term in our CQP optimization problem in Eq. (4.5), we can tune and define the importance which we grant for the goal of lack of complexity and, by this, for the antagonistic goal of accuracy.

4.2 Numerical Experience on a Complex Multi-Model Regulatory Networks

4.2.1 Data Description

To exemplify the implementation of MARS and CMARS algorithms, we use an artificial dataset which has two targets and two environmental variables and we have four predictor variables $(\widetilde{x}_1, \widetilde{x}_2, \widetilde{e}_1, \widetilde{e}_2)$ with 25 measurement values for all target and environmental variables. For MARS and CMARS algorithm, first, the MARS models are constructed for each targets and environmental variable by using the Salford MARS [83] and, then, the maximum number of BFs ($M_{\max}$) and the highest degree of interactions are defined.

For the first target, $\hat{x}_1$, $M_{\max}$ is assigned to be 11, and the highest degree of interaction is assigned to be 1 which is the main model. Therefore, and provided the knot values (through MARS software), the largest model involves the subsequent BFs (for simplicity, we suppress the arguments of the model functions):

$$\vartheta_1 = \max\{0, \widetilde{e}_2 + 2.045\}, \qquad \vartheta_2 = \max\{0, \widetilde{e}_1 + 2.056\},$$
$$\vartheta_3 = \max\{0, \widetilde{x}_1 + 2.280\}, \qquad \vartheta_4 = \max\{0, \widetilde{x}_2 - 0.029\},$$
$$\vartheta_5 = \max\{0, 0.029 - \widetilde{x}_2\}, \qquad \vartheta_6 = \max\{0, \widetilde{x}_1 + 0.293\},$$
$$\vartheta_7 = \max\{0, -0.293 - \widetilde{x}_1\}, \qquad \vartheta_8 = \max\{0, \widetilde{e}_2 + 0.093\},$$
$$\vartheta_9 = \max\{0, -0.093 - \widetilde{e}_2\}, \qquad \vartheta_{10} = \max\{0, \widetilde{e}_1 + 0.186\},$$
$$\vartheta_{11} = \max\{0, -0.186 - \widetilde{e}_1\}.$$

For the second target, $\hat{x}_2, M_{\max}$ is assigned to be 10, and the highest degree of interaction is assigned to be 1. So, the largest model includes the following BFs:

$$\vartheta_1 = \max\{0, \widetilde{e}_1 + 2.056\}, \qquad \vartheta_2 = \max\{0, \widetilde{e}_2 - 0.386\},$$
$$\vartheta_3 = \max\{0, 0.386 - \widetilde{e}_2\}, \qquad \vartheta_4 = \max\{0, \widetilde{x}_1 + 1.791\},$$
$$\vartheta_5 = \max\{0, \widetilde{x}_1 + 0.293\}, \qquad \vartheta_6 = \max\{0, -0.293 - \widetilde{x}_1\},$$
$$\vartheta_7 = \max\{0, \widetilde{x}_2 - 0.029\}, \qquad \vartheta_8 = \max\{0, 0.029 - \widetilde{x}_2\},$$
$$\vartheta_9 = \max\{0, \widetilde{x}_2 + 0.332\}, \qquad \vartheta_{10} = \max\{0, -0.332 - \widetilde{x}_2\}.$$

For the first environmental factor, $\hat{e}_1, M_{\max}$ and the highest degree of interaction is assigned to be 7 and 2, respectively. Consequently, the BFs of the largest model are represented as

$$\vartheta_1 = \max\{0, \widetilde{x}_1 + 2.280\}, \qquad \vartheta_2 = \max\{0, \widetilde{e}_1 + 2.056\},$$
$$\vartheta_3 = \max\{0, \widetilde{x}_2 + 1.791\}, \qquad \vartheta_4 = \max\{0, \widetilde{e}_2 + 0.017\} \cdot \vartheta_2,$$
$$\vartheta_5 = \max\{0, -0.017 - \widetilde{e}_2\} \cdot \vartheta_2, \qquad \vartheta_6 = \max\{0, \widetilde{x}_1 + 0.293\} \cdot \vartheta_3,$$
$$\vartheta_7 = \max\{0, -0.293 - \widetilde{x}_1\} \cdot \vartheta_3.$$

For the second environmental factor, $\hat{e}_2$, $M_{\max}$, is assigned to be 11, and the highest degree of interaction is assigned to be 1. Therefore, the largest model involves the following BFs:

$$\vartheta_1 = \max\{0, \widetilde{e}_2 + 2.045\}, \qquad \vartheta_2 = \max\{0, \widetilde{e}_1 - 0.443\},$$
$$\vartheta_3 = \max\{0, 0.443 - \widetilde{e}_1\}, \qquad \vartheta_4 = \max\{0, \widetilde{x}_1 + 0.293\},$$
$$\vartheta_5 = \max\{0, -0.293 - \widetilde{x}_1\}, \qquad \vartheta_6 = \max\{0, \widetilde{x}_2 - 0.029\},$$
$$\vartheta_7 = \max\{0, 0.029 - \widetilde{x}_2\}, \qquad \vartheta_8 = \max\{0, \widetilde{e}_1 + 0.186\},$$
$$\vartheta_9 = \max\{0, -0.186 - \widetilde{e}_1\}, \qquad \vartheta_{10} = \max\{0, \widetilde{x}_2 + 0.332\},$$
$$\vartheta_{11} = \max\{0, -0.332 - \widetilde{x}_2\}.$$

For all target and environmental variables, using these BFs above, the largest models with M_{max} BFs and the final (optimally estimated) models with the reduced number of BFs are constructed after the forward and the backward step of MARS by its software. At the end, the final models used for MARS algorithm and the largest models used for CMARS algorithm are found and represented in Sects. 4.2.2 and 4.2.3, respectively.

4.2.2 MARS Models

After the backward stepwise elimination of MARS, for both targets and environmental factors, the numbers of BFs are reduced to 5, 5, 5 and 6, respectively. Consequently, for this study, the final models of MARS are obtained in the subsequent form of estimations:

$$\begin{aligned}\hat{x}_1 = {} & \alpha_0 + \alpha_1 \max\{0, \widetilde{e}_2 + 2.045\} + \alpha_2 \max\{0, \widetilde{e}_1 + 2.056\} \\ & + \alpha_3 \max\{0, \widetilde{x}_1 + 2.280\} + \alpha_4 \max\{0, \widetilde{x}_2 + 0.029\} \\ & + \alpha_5 \max\{0, -0.029 - \widetilde{x}_2\},\end{aligned}$$

$$\begin{aligned}\hat{x}_2 = {} & \alpha_0 + \alpha_1 \max\{0, \widetilde{e}_1 + 2.056\} + \alpha_2 \max\{0, \widetilde{e}_2 - 0.386\} \\ & + \alpha_3 \max\{0, 0.386 - \widetilde{e}_2\} + \alpha_4 \max\{0, -0.293 - \widetilde{x}_1\} \\ & + \alpha_5 \max\{0, \widetilde{x}_2 + 0.029\},\end{aligned}$$

Table 4.1 For targets and environmental factors, parameter values of MARS algorithm

	α_0	α_1	α_2	α_3	α_4	α_5	α_6	α_7
$\widetilde{x}_1$	−0.452	0.298	−0.959	0.788	−0.152	0.184		
$\widetilde{x}_2$	1.135	−0.626	−0.859	0.548			0.181	−0.206
$\widetilde{e}_1$	−3.939	0.749	0.764	0.360		−0.096		0.155
$\widetilde{e}_2$	−2.134	0.672	−0.448	1.087	0.634	−0.252	−0.369	

$$\begin{aligned}\hat{e}_1 &= \alpha_0 + \alpha_1 \max\{0, \widetilde{x}_1 + 2.280\} + \alpha_2 \max\{0, \widetilde{e}_1 + 2.056\} \\ &\quad + \alpha_3 \max\{0, \widetilde{x}_2 + 1.791\} + \alpha_4 \max\{0, -0.017 - \widetilde{e}_2\} \cdot \max\{0, \widetilde{e}_1 + 2.056\} \\ &\quad + \alpha_5 \max\{0, -0.293 - \widetilde{x}_1\} \cdot \max\{0, \widetilde{x}_2 + 1.791\},\end{aligned}$$

$$\begin{aligned}\hat{e}_2 &= \alpha_0 + \alpha_1 \max\{0, \widetilde{e}_2 + 2.045\} + \alpha_2 \max\{0, \widetilde{e}_1 - 0.443\} \\ &\quad + \alpha_3 \max\{0, 0.443 - \widetilde{e}_1\} + \alpha_4 \max\{0, \widetilde{x}_1 + 0.293\} \\ &\quad + \alpha_5 \max\{0, -0.293 - \widetilde{x}_1\} + \alpha_6 \max\{0, \widetilde{x}_2 - 0.029\}.\end{aligned}$$

For each target and environmental factor, the unknown parameters are determined and represented in Table 4.1.

4.2.3 *CMARS Models*

For CMARS algorithm, to prevent from nondifferentiability in our optimization problem, we choose the knot values different from data points, but very much nearby to the corresponding input data. For the first part of our optimization problem in Eq. (4.4), using M_{max} BFs represented in Sect. 4.2.1, the largest models become

$$\begin{aligned}\hat{x}_1 &= \alpha_0 + \alpha_1 \max\{0, \widetilde{e}_2 + 2.046\} + \alpha_2 \max\{0, \widetilde{e}_1 + 2.057\} \\ &\quad + \alpha_3 \max\{0, \widetilde{x}_1 + 2.281\} + \alpha_4 \max\{0, \widetilde{x}_2 + 0.030\} \\ &\quad + \alpha_5 \max\{0, -0.030 - \widetilde{x}_2\} + \alpha_6 \max\{0, \widetilde{x}_1 + 0.294\} \\ &\quad + \alpha_7 \max\{0, 0.030 - \widetilde{x}_2\} + \alpha_8 \max\{0, \widetilde{e}_1 + 0.186\} \\ &\quad + \alpha_9 \max\{0, -0.186 - \widetilde{e}_1\} + \alpha_{10} \max\{0, \widetilde{x}_2 + 0.333\} \\ &\quad + \alpha_{11} \max\{0, -0.186 - \widetilde{e}_1\},\end{aligned}$$

$$\begin{aligned}
\hat{x}_2 = {} & \alpha_0 + \alpha_1 \max\{0, \widetilde{e}_1 + 2.057\} + \alpha_2 \max\{0, \widetilde{e}_2 - 0.387\} \\
& + \alpha_3 \max\{0, 0.387 - \widetilde{e}_2\} + \alpha_4 \max\{0, \widetilde{x}_1 + 1.792\} \\
& + \alpha_5 \max\{0, \widetilde{x}_1 + 0.294\} + \alpha_6 \max\{0, -0.294 - \widetilde{x}_1\} + \\
& + \alpha_7 \max\{0, \widetilde{x}_2 + 0.030\} + \alpha_8 \max\{0, 0.030 - \widetilde{x}_2\} \\
& + \alpha_9 \max\{0, \widetilde{x}_2 + 0.333\} + \alpha_{10} \max\{0, -0.333 - \widetilde{x}_2\},
\end{aligned}$$

$$\begin{aligned}
\hat{e}_1 = {} & \alpha_0 + \alpha_1 \max\{0, \widetilde{x}_1 + 2.281\} + \alpha_2 \max\{0, \widetilde{e}_1 + 2.057\} \\
& + \alpha_3 \max\{0, \widetilde{x}_2 + 1.792\} + \alpha_4 \max\{0, \widetilde{e}_2 + 0.018\} \cdot \max\{0, \widetilde{e}_1 + 2.057\} \\
& + \alpha_5 \max\{0, -0.018 - \widetilde{e}_2\} \cdot \max\{0, \widetilde{e}_1 + 2.057\} \\
& + \alpha_6 \max\{0, \widetilde{x}_1 + 0.294\} \cdot \max\{0, \widetilde{x}_2 + 1.792\} \\
& + \alpha_7 \max\{0, -0.294 - \widetilde{x}_1\} \cdot \max\{0, \widetilde{x}_2 + 1.792\},
\end{aligned}$$

$$\begin{aligned}
\hat{e}_2 = {} & \alpha_0 + \alpha_1 \max\{0, \widetilde{e}_2 + 2.046\} + \alpha_2 \max\{0, \widetilde{e}_1 - 0.444\} \\
& + \alpha_3 \max\{0, 0.444 - \widetilde{e}_1\} + \alpha_4 \max\{0, \widetilde{x}_1 + 0.294\} \\
& + \alpha_5 \max\{0, -0.294 - \widetilde{x}_1\} + \alpha_6 \max\{0, \widetilde{x}_2 - 0.030\} \\
& + \alpha_7 \max\{0, 0.030 - \widetilde{x}_2\} + \alpha_8 \max\{0, \widetilde{e}_1 + 0.187\} \\
& + \alpha_9 \max\{0, -0.187 - \widetilde{e}_1\} + \alpha_{10} \max\{0, \widetilde{x}_2 + 0.333\} \\
& + \alpha_{10} \max\{0, -0.333 - \widetilde{x}_2\}.
\end{aligned}$$

After the discretized form of multi-dimensional integrals in Eq. (3.5) is denoted by $\boldsymbol{L}$, for the second part of our optimization model in Eq. (4.4), the $\boldsymbol{L}$ matrices of each target and each environmental factor become diagonal (12×12)-, (11×11)-, (8×8)- and (12×12)-matrices, and the first column elements of $\boldsymbol{L}$ are all zero. For instance, the $\boldsymbol{L}$ matrix of first environmental item can be presented as follows:

$$\boldsymbol{L} = \begin{bmatrix} 0 & 0 & \dots & 0 \\ 0 & 1.9671 & \dots & 0 \\ \vdots & \vdots & \ddots & \vdots \\ 0 & 0 & \dots & 1.3287 \end{bmatrix},$$

and $\|\boldsymbol{L\alpha}\|_2^2$ is given as

$$\begin{aligned}\|\boldsymbol{L\alpha}\|_2^2 =& (1.967\cdot\alpha_1)^2 + (1.972\cdot\alpha_2)^2 + (1.911\cdot\alpha_3)^2\\ &+ (1.401\cdot\alpha_4)^2 + (1.381\cdot\alpha_5)^2 + (1.502\cdot\alpha_6)^2\\ &+ (1.329\cdot\alpha_7)^2.\end{aligned}$$

After we obtain largest models for the accuracy part and evaluate the $\boldsymbol{L}$ matrices for complexity part of PRSS in Eq. (4.4) for the first environmental factor, we reformulate PRSS as a problem of CQP by using Eq. (4.5) as follows:

$$\begin{aligned}&\underset{w,\boldsymbol{\alpha}}{\text{minimize}}\ w,\\ &\text{subject to}\\ &1.4624-\alpha_0-3.788\alpha_1-3.461\alpha_2-1.165\alpha_3-5.114\alpha_5-2.099\alpha_6=\beta_1,\\ &0.3915-\alpha_0-1.737\alpha_1-3.384\alpha_2-1.821\alpha_3-3.309\alpha_5-0.455\alpha_7=\beta_2,\\ &\vdots\\ &-0.637-\alpha_0-1.778\alpha_1-1.877\alpha_2-1.548\alpha_3-0.883\alpha_5-0.332\alpha_7=\beta_{25},\\ &(\beta_1^2+\beta_2^2+\ldots+\beta_{25}^2)^{1/2}\le w,\\ &(\beta_{26}^2+\beta_{27}^2+\beta_{28}^2+\beta_{29}^2+\beta_{30}^2+\beta_{31}^2+\beta_{32}^2)^{1/2}\le(K)^{1/2}.\end{aligned}$$

Here, we recall that the values K are determined by a model-free (train and error) method (cf. Remark 1). After solving this problem for all target and environmental factors, we receive unknown parameters which are presented as given in Table 4.2. For all computations, the code written in MATLAB is run MOSEK software [87] is used for CQP.

Table 4.2 For targets and environmental factors, parameter values of CMARS algorithm

	α_0	α_1	α_2	α_3	α_4	α_5
$\tilde{x}_1$	−0.373	0.127	−0.108	0.119	0.193	−0.059
$\tilde{x}_2$	0.268	−0.230	−0.389	0.229	0.112	−0.241
$\tilde{e}_1$	−0.506	0.084	0.068	−0.027	0.168	0.019
$\tilde{e}_2$	−0.801	0.273	−0.090	0.099	0.247	−0.153
	α_6	α_7	α_8	α_9	α_{10}	α_{11}
$\tilde{x}_1$	0.129	−0.104	0.122	−0.125	−0.134	0.079
$\tilde{x}_2$	0.122	0.145	−0.086	0.109	−0.110	
$\tilde{e}_1$	0.117	−0.153				
$\tilde{e}_2$	0.203	−0.052	−0.120	−0.075	0.153	−0.047

4.2.4 Results and Comparison

The prediction results for targets and environmental factors can be seen in Figs. 4.1, 4.2, 4.3, 4.4, where 'blue line' present real values, 'red line' indicates the estimated values by MARS model and 'green line' represents the predicted values by CMARS model.

As we may deduce from Figs. 4.1–4.4, with the real expression values of targets and environmental factors, the predicted values of CMARS model match much better than that of MARS model. In fact, this indicates that CMARS can

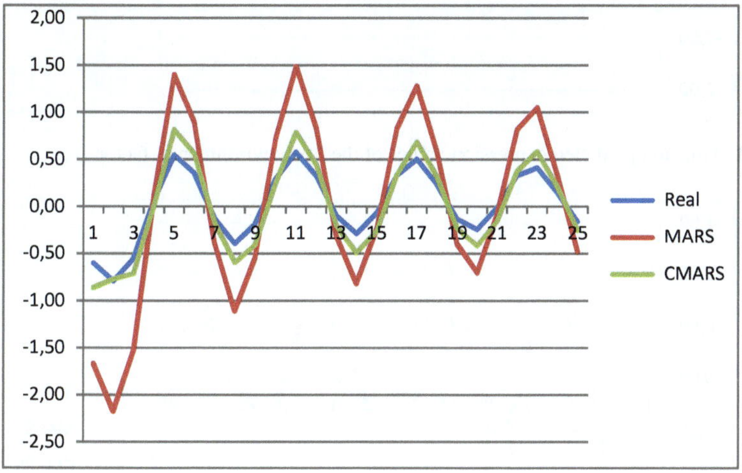

Fig. 4.1 True and predicted expression values of the first target

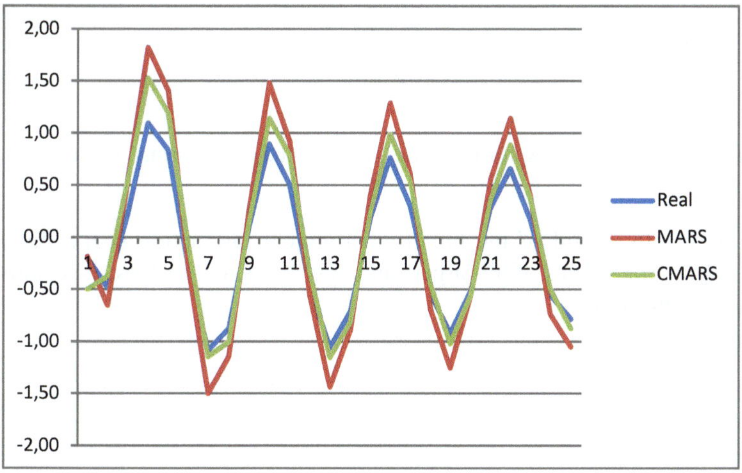

Fig. 4.2 True and predicted expression values of the second target

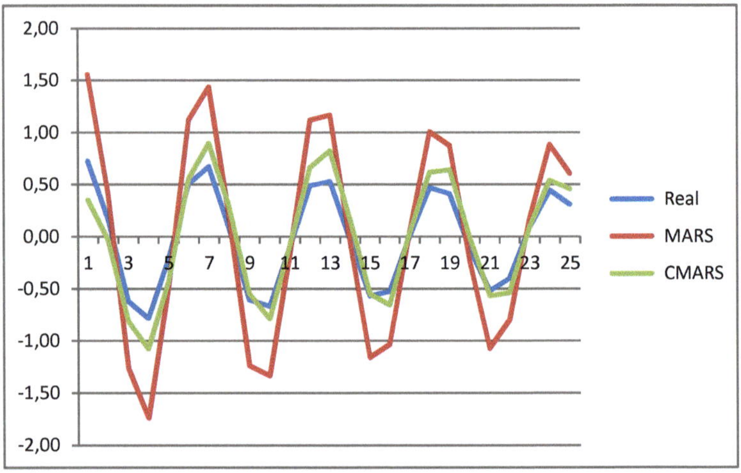

Fig. 4.3 True and predicted expression values of the first environmental factor

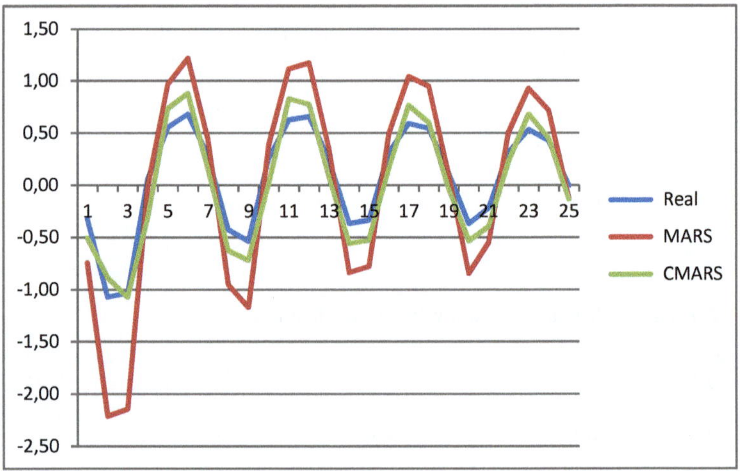

Fig. 4.4 True and predicted expression values of the second environmental factor

really predict the trend of the target-environment interaction successfully based on the expression values of all targets and environmental factors, especially, when compared with MARS.

4.3 Simulation Study

In previous sections, we represented and investigated scientifically MARS, CMARS and two-model regulatory systems with spline entries. In previous subsections, using an artificial data set, we introduced MARS and CMARS models for 2 target and 2 environmental factors as a numerical example and presented the results obtaining figures for each target and each environmental item. Now, in order to show the performance of MARS and CMARS for regulatory system based on replicated datasets, we constructed different MARS and CMARS models through 5 different simulated datasets for each of the target and environmental items, as we described in previous Sects. 4.2.1–4.2.3. Therefore, unknown parameters are determined and presented in Tables A.1 and A.2. Afterwards, these models are evaluated with respect to the criteria by using the formulas as given in Table D.1. To compare the results concerning the accuracy of MARS and CMARS, the models are calculated based on the adjusted multiple coefficients of determination (R^2_{adj}), average absolute error (*AAE*), root mean squared error (*RMSE*), and the correlation coefficient (r). The explanations, interpretations and formulas of these measures are represented in Table D.1, and the results are displayed in Table A.3.

According to these accuracy criteria, we understand that CMARS can perform better than MARS for all target and environmental items with respect to all measures validated through simulated datasets.

In spite of the recorded successes, the statistical methods like MARS and CMARS, which assume the input data are usually known precisely in developing models, may not give trustworthy results since, in reality, the data involved in regression problems can contain noise. Therefore, it has been realized that core elements of a new global regulatory framework have to be created to make these systems more robust and suitable for serving the requirements of the real world. In order to reveal this expectation, in the following chapter, a new robust optimization technique for solving and optimizing models implying nonlinearity and uncertainty by using R(C)MARS is presented with an implementation on two-factor regulatory systems. This will allow us to involve into our modeling *uncertainty* in the input variables, which is typical for so many real-life problems.

Chapter 5
Robust Optimization in Spline Regression Models for Regulatory Networks Under Polyhedral Uncertainty

In the previous chapter, we introduced and investigated new dynamical regression problems by using splines for the entries of regulatory network, and we demonstrated the effectiveness of these approaches by a numerical experiment. For that study [103], CMARS provides better results than MARS and gives us better predictions than MARS of the trend of the TE interaction based on the expression values of all targets and all environmental factors. These systems appear in the financial sector, in banking, environmental protection, system biology, medicine, etc. As practitioners in these fields need to be aware that the evaluation of probabilities based on history could be fundamentally inaccurate, uncertainties have a great importance for actors in these sectors. Therefore, in this chapter, our new robust optimization technique for solving and optimizing models having nonlinearity and uncertainty by using R(C)MARS is discussed with an implementation on two-factor TE systems [106].

5.1 Robustification of Regression for Regulatory Networks

Identification of a regulatory network from given real-world data is a mathematical problem that has to be solved both theoretically and computationally, especially, if there exists noise in the data. Given this motivation we discuss and newly present a robustification of regression problems for time-discrete TE regulatory systems under polyhedral uncertainty by using RCMARS. In our considered case of uncertainty existing in all kinds of the expression data, where the uncertainty sets are defined in Eqs. (5.17)–(5.18), RCMARS is applied to guarantee a robustification of our target-environment networks.

For RCMARS, the large model that has the maximum number of BFs, M_{max}, is created by Salford MARS [83]. In that process, the input and output variables of our

A. Özmen, *Robust Optimization of Spline Models and Complex Regulatory Networks*, Contributions to Management Science,
DOI 10.1007/978-3-319-30800-5_5

model are all assumed as random variables for target-environment networks. They lead us to uncertainty sets; those are assumed to contain CIs. Furthermore, $\widetilde{X}_j, \widetilde{E}_i$ and, in vector form: $\widehat{\boldsymbol{X}}$ and $\widehat{\boldsymbol{E}}$, are considered to be normally distributed. So, the following general model is considered for input data values, $\widetilde{X}_j$ and $\widetilde{E}_i$:

$$\begin{aligned} \widetilde{X}_j &= \bar{\widetilde{x}}_j + \zeta_j^{\mathrm{T}} \quad (j = 1, 2, \ldots, n), \\ \widetilde{E}_i &= \bar{\widetilde{e}}_i + \zeta_i^{\mathrm{E}} \quad (i = 1, 2, \ldots, m). \end{aligned} \tag{5.1}$$

Here, $\bar{\widetilde{x}}_j$ and $\bar{\widetilde{e}}_i$ denote the sample mean (average) of the input vectors $\widetilde{\boldsymbol{X}}_j$ and $\widetilde{\boldsymbol{E}}_i$, respectively. When considering that we have $d\,(= n + m)$-dimensional input data, each input vector $\widetilde{\boldsymbol{x}}_k = (\widetilde{x}_{k,1}, \widetilde{x}_{k,2}, \ldots, \widetilde{x}_{k,n})^T$ for target and each input vector $\widetilde{\boldsymbol{e}}_k = (\widetilde{e}_{k,1}, \widetilde{e}_{k,2}, \ldots, \widetilde{e}_{k,m})^T$ for environment are represented as $\widetilde{\widetilde{\boldsymbol{x}}}_k = (\widetilde{\widetilde{x}}_{k,1}, \widetilde{\widetilde{x}}_{k,2}, \ldots, \widetilde{\widetilde{x}}_{k,n})^T$ and $\widetilde{\widetilde{\boldsymbol{e}}}_k = (\widetilde{\widetilde{e}}_{k,1}, \widetilde{\widetilde{e}}_{k,2}, \ldots, \widetilde{\widetilde{e}}_{k,m})^T$ including the perturbations $\boldsymbol{\Delta}_k^{\mathrm{T}} = (\Delta_{k,1}^{\mathrm{T}}, \ldots, \Delta_{k,n}^{\mathrm{T}})^T$ and $\boldsymbol{\Delta}_k^{\mathrm{E}} = (\Delta_{k,1}^{\mathrm{E}}, \ldots, \Delta_{k,m}^{\mathrm{E}})^T$, respectively $(k = 0, 1, \ldots, N)$. Here, $\boldsymbol{\Delta}_k^{\mathrm{T}}$ and $\boldsymbol{\Delta}_k^{\mathrm{E}}$ are generic elements of U_1^{T} and U_1^{E}, which are the polyhedral uncertainty sets that will later on be described for our input data (cf. Eq. (5.17)). So, for TE networks, the new values of piecewise linear BFs are represented as follows:

$$\begin{aligned} &\widetilde{x}_{k,j} \to \widetilde{\widetilde{x}}_{k,j}; \quad \widetilde{\widetilde{x}}_{k,j} = \bar{\widetilde{x}}_j + \Delta_{k,j}^{\mathrm{T}}, \quad |\Delta_{k,j}^{\mathrm{T}}| \le \rho_{k,j}^{\mathrm{T}} \;\; (j = 1, 2, \ldots, n; k = 0, 1, \ldots, N), \\ &e_{k,i} \to \widetilde{\widetilde{e}}_{k,i}; \quad \widetilde{\widetilde{e}}_{k,i} = \bar{\widetilde{e}}_i + \Delta_{k,i}^{\mathrm{E}}, \quad |\Delta_{k,i}^{\mathrm{E}}| \le \rho_{k,i}^{\mathrm{E}} \;\; (i = 1, 2, \ldots, m; k = 0, 1, \ldots, N), \end{aligned}$$

Similarly, after we incorporate a perturbation into output variables, the output vectors $\hat{\boldsymbol{x}} = (\hat{x}_1, \hat{x}_2, \ldots, \hat{x}_n)^T$ for target and $\hat{\boldsymbol{e}} = (\hat{e}_1, \hat{e}_2, \ldots, \hat{e}_m)^T$ for environment are represented as $\hat{\hat{\boldsymbol{x}}} = (\hat{\hat{x}}_1, \hat{\hat{x}}_2, \ldots, \hat{\hat{x}}_n)^T$ and $\hat{\hat{\boldsymbol{e}}} = (\hat{\hat{e}}_k, \hat{\hat{e}}_k, \ldots, \hat{\hat{e}}_k)^T$ including the perturbations $\boldsymbol{\tau}^{\mathrm{T}} = (\tau_1^{\mathrm{T}}, \tau_2^{\mathrm{T}}, \ldots, \tau_n^{\mathrm{T}})^T$ and $\boldsymbol{\tau}^{\mathrm{E}} = (\tau_1^{\mathrm{E}}, \tau_2^{\mathrm{E}}, \ldots, \tau_m^{\mathrm{E}})^T$, respectively. Here, we restrict the vectors $\boldsymbol{\tau}^{\mathrm{T}}$ and $\boldsymbol{\tau}^{\mathrm{E}}$ to be elements of U_1^{T} and U_1^{E}, which are the polyhedral uncertainty sets that will later on be defined for our output data (cf. Eq. (5.18)). So, our new output values can be expressed as follows:

$$\begin{aligned} &\hat{x}_j \to \hat{\hat{x}}_j; \quad \hat{\hat{x}}_j = \bar{\hat{x}}_j + \tau_j^{\mathrm{T}}, \quad |\tau_j^{\mathrm{T}}| \le \nu_j^{\mathrm{T}} \;\; (j = 1, 2, \ldots, n;\; k = 0, 1, \ldots, N), \\ &\hat{e}_i \to \hat{\hat{e}}_i; \quad \hat{\hat{e}}_i = \bar{\hat{e}}_i + \tau_i^{\mathrm{E}}, \quad |\tau_i^{\mathrm{E}}| \le \nu_i^{\mathrm{E}} \;\; (i = 1, 2, \ldots, m;\; k = 0, 1, \ldots, N), \end{aligned}$$

where $\bar{\hat{x}}_j$ and $\bar{\hat{e}}_i$ express the sample mean (average) of the output vectors $\hat{\boldsymbol{X}}$ and $\hat{\boldsymbol{E}}$, respectively. When we estimate the BFs in Eq. (3.13) with uncertainty for TE, we can evaluate them through the subsequent estimations:

$$\begin{aligned} &\left[\widetilde{\widetilde{x}}_{k,\upsilon(j,n)} - \tau_{\upsilon(j,n)}^{\mathrm{T}}\right]_{\pm} \le \left[\widetilde{x}_{k,\upsilon(j,n)} - \tau_{\upsilon(j,n)}^{\mathrm{T}}\right]_{\pm} + \left[\Delta_{k,\upsilon(j,n)}^{\mathrm{T}} + (\pm \mathrm{A}_{k,\upsilon(j,n)}^{\mathrm{T}})\right]_{\pm}, \\ &\left[\widetilde{\widetilde{e}}_{k,\upsilon(i,m)} - \tau_{\upsilon(i,m)}^{\mathrm{E}}\right]_{\pm} \le \left[\widetilde{e}_{k,\upsilon(i,m)} - \tau_{\upsilon(i,m)}^{\mathrm{E}}\right]_{\pm} + \left[\Delta_{k,\upsilon(i,m)}^{\mathrm{E}} + (\pm \mathrm{A}_{k,\upsilon(i,m)}^{\mathrm{E}})\right]_{\pm}; \end{aligned} \tag{5.2}$$

here, $\mathrm{A}^{\mathrm{T}}_{k,v(j,n)}$ and $\mathrm{A}^{\mathrm{E}}_{k,v(i,m)}$ are interpreted and employed as *control variables*.[1] Since the values of these control variable directly influence the size of our uncertainty set U_1, and our uncertainty sets are unknown but bounded, $\mathrm{A}^{\mathrm{T}}_{k,v(j,n)}$ and $\mathrm{A}^{\mathrm{E}}_{k,v(i,m)}$ are restricted by values $\gamma^{\mathrm{T}}_{k,v(j,n)}$ and $\gamma^{\mathrm{E}}_{k,v(i,m)}$, respectively. If we encounter the very conservative (risk-averse) position, the so-called *worst case* for the values of $\mathrm{A}^{\mathrm{T}}_{k,v(j,n)}$ and $\mathrm{A}^{\mathrm{E}}_{k,v(i,m)}$, they will be equal to $\gamma^{\mathrm{T}}_{k,v(j,n)}$ and $\gamma^{\mathrm{E}}_{k,v(i,m)}$, respectively. But if the absolute value of our uncertainty is very high, we might not find any meaningful solution for our problems. For this reason, we may allow for a more risk-friendly case by selecting the values of $\mathrm{A}^{\mathrm{T}}_{k,v(j,n)}$ and $\mathrm{A}^{\mathrm{E}}_{k,v(i,m)}$ between 0 and the absolute value of $\mathrm{A}^{\mathrm{T}}_{k,v(j,n)}$ and $\mathrm{A}^{\mathrm{E}}_{k,v(i,m)}$. This means: $\breve{\mathrm{A}}^{\mathrm{T}}_{k,v(j,n)} \in [0, \left|\mathrm{A}^{\mathrm{T}}_{k,v(j,n)}\right|]$ and $\breve{\mathrm{A}}^{\mathrm{E}}_{k,v(i,m)} \in [0, \left|\mathrm{A}^{\mathrm{E}}_{k,v(i,m)}\right|]$, respectively. To make our notation a bit easier, we still keep the names $\mathrm{A}^{\mathrm{T}}_{k,v(j,n)}$ and $\mathrm{A}^{\mathrm{E}}_{k,v(i,m)}$ for $\breve{\mathrm{A}}^{\mathrm{T}}_{k,v(j,n)}$ and $\breve{\mathrm{A}}^{\mathrm{E}}_{k,v(i,m)}$. Now, to evaluate the values and differences of $\vartheta_n(\widetilde{\widetilde{\boldsymbol{x}}}^n_k)$ and $\vartheta_n(\widetilde{\boldsymbol{x}}^n_k)$ for the targets, $\vartheta_m(\widetilde{\widetilde{\boldsymbol{e}}}^m_k)$ and $\vartheta_m(\widetilde{\boldsymbol{e}}^m_k)$ for the environmental items in Eq. (3.13), we can apply Eq. (5.2) in the following way, where all the '+' and '−' signs correspond to each other, respectively: the values of these control variable directly influence the size of our uncertainty set U^{T}_1 and U^{E}_1, $\mathrm{A}^{\mathrm{T}}_{k,v(j,n)}$ and $\mathrm{A}^{\mathrm{E}}_{k,v(i,m)}$ are restricted by values $\gamma^{\mathrm{T}}_{k,v(j,n)}$ and $\gamma^{\mathrm{E}}_{k,v(i,m)}$, respectively:

$$\underbrace{\prod_{j=1}^{K_n}[\widetilde{\widetilde{x}}_{k,v(j,n)} - \varphi_{v(j,n)}]_\pm}_{=:\vartheta_n(\widetilde{\widetilde{\boldsymbol{x}}}^n_k)} \leq \underbrace{\prod_{j=1}^{K_n}[\widetilde{x}_{k,v(j,n)} - \varphi_{v(j,n)}]_\pm}_{=:\vartheta_n(\widetilde{\boldsymbol{x}}^n_k)} + \sum_{\substack{A \subsetneq \{1,\ldots,K_n\}}} \prod_{a \in A}[\widetilde{\widetilde{x}}_{k,a} - \varphi^{\mathrm{T}}_a]_\pm \prod_{b \in \{1,\ldots,K_n\}/A} [(\pm \mathrm{A}^{\mathrm{T}}_{k,b}) + \Delta^{\mathrm{T}}_{k,b}]_\pm, \tag{5.3}$$

$$\underbrace{\prod_{i=1}^{K_m}[\widetilde{\widetilde{e}}_{k,v(i,m)} - \varphi_{v(i,m)}]_\pm}_{=:\vartheta_m(\widetilde{\widetilde{\boldsymbol{e}}}^m_k)} \leq \underbrace{\prod_{i=1}^{K_m}[\widetilde{e}_{k,v(i,m)} - \varphi_{v(i,m)}]_\pm}_{=:\vartheta_m(\widetilde{\boldsymbol{e}}^m_k)} + \sum_{\substack{A \subsetneq \{1,\ldots,K_m\}}} \prod_{a \in A}[\widetilde{\widetilde{e}}_{k,a} - \varphi^{\mathrm{E}}_a]_\pm \prod_{b \in \{1,\ldots,K_m\}/A} [(\pm \mathrm{A}^{\mathrm{E}}_{k,b}) + \Delta^{\mathrm{E}}_{k,b}]_\pm. \tag{5.4}$$

[1]There should be no confusion by double use of the letters n and m for both number of variables and dimension of subvectors in R(C)MARS model.

Here, we may achieve a bounding given below via symmetry, namely:

$$\left.\begin{array}{l}\vartheta_n(\widetilde{\widetilde{x}}_k^n) - \vartheta_n(\widetilde{x}_k^n) \leq \hat{u}_{k,n}^{\mathrm{T}}, \\ \vartheta_n(\widetilde{x}_k^n) - \vartheta_n(\widetilde{\widetilde{x}}_k^n) \leq \hat{\hat{u}}_{k,n}^{\mathrm{T}}\end{array}\right\} \Rightarrow \left|\vartheta_n(\widetilde{\widetilde{x}}_k^n) - \vartheta_n(\widetilde{x}_k^n)\right| \leq \max\left\{\hat{u}_{k,n}^{\mathrm{T}}, \hat{\hat{u}}_{k,n}^{\mathrm{T}}\right\}, \tag{5.5}$$

$$\left.\begin{array}{l}\vartheta_m(\widetilde{\widetilde{e}}_k^m) - \vartheta_m(\widetilde{e}_k^m) \leq \hat{u}_{k,m}^{\mathrm{E}} \\ \vartheta_m(\widetilde{e}_k^m) - \vartheta_m(\widetilde{\widetilde{e}}_k^m) \leq \hat{\hat{u}}_{k,m}^{\mathrm{E}}\end{array}\right\} \Rightarrow \left|\vartheta_m(\widetilde{\widetilde{e}}_k^m) - \vartheta_m(\widetilde{e}_k^m)\right| \leq \max\left\{\hat{u}_{k,m}^{\mathrm{E}}, \hat{\hat{u}}_{k,m}^{\mathrm{E}}\right\}. \tag{5.6}$$

Therefore, our uncertainty values $\left|u_{kn}^{\mathrm{T}}\right|$ for target and $\left|u_{km}^{\mathrm{E}}\right|$ for environment can be estimated in the subsequent manner for every BF:

$$\begin{aligned}\left|u_{k,n}^{\mathrm{T}}\right| &\leq \sum_{\substack{A \subseteq \{1,\ldots,K_n\} \\ \neq}} \mathrm{B}_{k\mathrm{T}}^{|A|-1} \prod_{a \in A} \rho_{k,a}^{\mathrm{T}} \cdot \prod_{b \in \{1,\ldots,K_n\}/A} (\gamma_{k,b}^{\mathrm{T}} + \rho_{k,b}^{\mathrm{T}}), \\ \left|u_{k,m}^{\mathrm{E}}\right| &\leq \sum_{\substack{A \subseteq \{1,\ldots,K_m\} \\ \neq}} \mathrm{B}_{k\mathrm{E}}^{|A|-1} \prod_{a \in A} \rho_{k,a}^{\mathrm{E}} \cdot \prod_{b \in \{1,\ldots,K_m\}/A} (\gamma_{k,b}^{\mathrm{E}} + \rho_{k,b}^{\mathrm{E}}).\end{aligned} \tag{5.7}$$

Here, by $|A|$ we imply the cardinality (size) of the set A. The values of $\mathrm{B}_{k\mathrm{T}}$ and $\mathrm{B}_{k\mathrm{E}}$ are regarded and applied again as *control variables*. The values of $\mathrm{B}_{k\mathrm{T}}$ and $\mathrm{B}_{k\mathrm{E}}$ are equal to 2 in cases without outliers, while, given outliers, they will be greater than 2. In those situations, we shall choose different values for $\mathrm{B}_{k\mathrm{T}}$ and $\mathrm{B}_{k\mathrm{E}}$. If we allow for a very conservative case, we do not wish to exclude any outliers. However, the values of $\mathrm{B}_{k\mathrm{T}}$ and $\mathrm{B}_{k\mathrm{E}}$ could be rather large for some variables in the input data, and the absolute values of our uncertainty sets might be quite high because of the values of these control variables. If the absolute value of any uncertainty set is very high, it can take too much time to catch a solution or we could not find a meaningful solution for our problem at all. For those reasons, rather than choosing a very conservative position, we may take into account a more risk-friendly position by choosing the values of $\mathrm{B}_{k\mathrm{T}}$ and $\mathrm{B}_{k\mathrm{E}}$ with a possible exclusion of the outliers. In our novel study, we would like to visualize the concept of robustification for the targets and environmental items by Figs. 5.1, 5.2, respectively.

After implying uncertainty into Eq. (4.1) of Sect. 4.1.2, we may state the following prediction equations:

$$\begin{aligned}\widehat{X}_j^{(k+1)} &= \alpha_{j0}^{\mathrm{T}} + \vartheta_{a_j\mathrm{TT}}(\widetilde{\widetilde{\boldsymbol{X}}}^{(k)}) + \vartheta_{a_j\mathrm{ET}}(\widetilde{\widetilde{\boldsymbol{E}}}^{(k)}), \\ \widehat{E}_i^{(k+1)} &= \alpha_{i0}^{\mathrm{E}} + \vartheta_{a_i\mathrm{TE}}(\widetilde{\widetilde{\boldsymbol{X}}}^{(k)}) + \vartheta_{a_j\mathrm{EE}}(\widetilde{\widetilde{\boldsymbol{E}}}^{(k)}).\end{aligned} \tag{5.8}$$

Hence, we can compare data and model predictions under uncertainty and obtain the following regression problem:

$$\text{minimize} \sum_{k=0}^{N} \left(\left\| \widehat{X}^{(k)} - \widetilde{\widetilde{X}}^{(k)} \right\|_2^2 + \left\| \widehat{E}^{(k)} - \widetilde{\widetilde{E}}^{(k)} \right\|_2^2 \right). \tag{5.9}$$

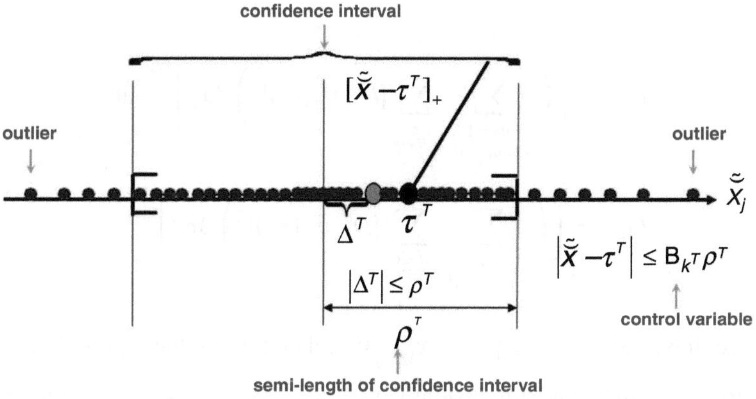

Fig. 5.1 The CIs for BF and perturbational term for target variables

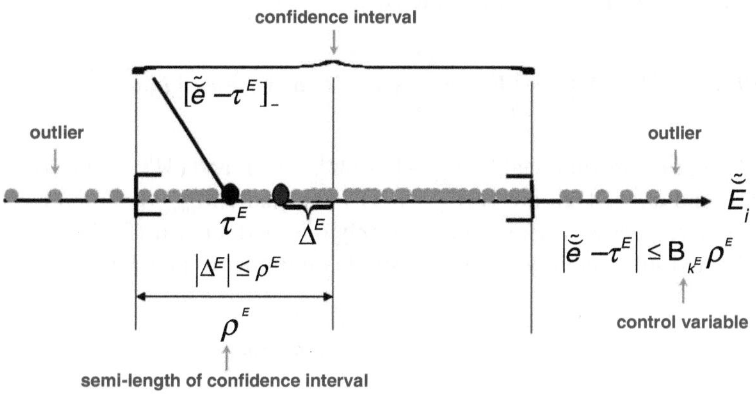

Fig. 5.2 The CIs for BF and perturbational term for environmental variables

Inserting splines and, then, including uncertainty expressed by *polyhedral uncertainty* sets and constructing the PRSS form of TE networks, the discretized form of PRSS of Eq. (3.5) attains the following expression:

$$PRSS \approx \sum_{k=0}^{N} \left(\sum_{j=1}^{n} (\tilde{\tilde{X}}_j^{(k)} - \tilde{\tilde{X}}_j^{(k)})^2 + \sum_{i=1}^{m} (\hat{\tilde{E}}_i^{(k)} - \hat{\tilde{E}}_i^{(k)})^2 \right)$$
$$+ \sum_{n=1}^{M_{\max}^{\mathrm{T}}} \phi_n \sum_{j=1}^{(N+1)^{K_n^{\mathrm{T}}}} L_{j,n}^2 \alpha_n^2 + \sum_{m=1}^{M_{\max}^{\mathrm{E}}} \phi_m \sum_{i=1}^{(N+1)^{K_m^{\mathrm{E}}}} L_{i,m}^2 \alpha_m^2, \tag{5.10}$$

where

$$L_{kn} := \left[\left(\sum_{\substack{|\theta|=1 \\ \theta=(\theta_1,\theta_2)^T}}^{2} \sum_{\substack{r<s \\ r,s\in V(n)}} [D^{\theta}_{r,s} \vartheta_n(\hat{x}^n_k)]^2 \right) \Delta \hat{x}^n_k \right]^{1/2} \text{ and}$$

$$L_{km} := \left[\left(\sum_{\substack{|\theta|=1 \\ \theta=(\theta_1,\theta_2)^T}}^{2} \sum_{\substack{r<s \\ r,s\in V(m)}} [D^{\theta}_{r,s} \vartheta_m(\hat{e}^m_k)]^2 \right) \Delta \hat{e}^m_k \right]^{1/2}.$$

Here, we have $\boldsymbol{\alpha}^{\mathrm{T}} := (\alpha^{\mathrm{T}}_0, \ldots, \alpha^{\mathrm{T}}_{M^{\mathrm{T}}_{\max}})^T$ related with the "point" (consisting of vectors of different dimensions) $\breve{\boldsymbol{x}}^{\mathrm{T}}_k := (\breve{\boldsymbol{x}}^1_k, \ldots, \breve{\boldsymbol{x}}^{M^{\mathrm{T}}_{\max}}_k)^T$, and $\boldsymbol{\alpha}^{\mathrm{E}} := (\alpha^{\mathrm{E}}_0, \alpha^{\mathrm{E}}_1, \ldots, \alpha^{\mathrm{E}}_{M^{\mathrm{E}}_{\max}})^T$ related with the "point" $\breve{\boldsymbol{e}}^{\mathrm{T}}_k := (\breve{\boldsymbol{e}}^1_k, \ldots, \breve{\boldsymbol{e}}^{M^{\mathrm{E}}_{\max}}_k)^{\mathrm{T}}$. Then, our approximation of PRSS may be written as:

$$PRSS \approx \left\| \widehat{\boldsymbol{X}} - \breve{\boldsymbol{X}} \right\|_2^2 + \left\| \widehat{\boldsymbol{E}} - \breve{\boldsymbol{E}} \right\|_2^2 + \phi_{\mathrm{T}} \left\| \boldsymbol{L}_{\mathrm{T}} \boldsymbol{\alpha}_{\mathrm{T}} \right\|_2^2 + \phi_{\mathrm{E}} \left\| \boldsymbol{L}_{\mathrm{E}} \boldsymbol{\alpha}_{\mathrm{E}} \right\|_2^2, \tag{5.11}$$

where $\boldsymbol{L}_{\mathrm{T}}$ and $\boldsymbol{L}_{\mathrm{E}}$ are diagonal $(M^{\mathrm{T}}_{\max}+1)\times(M^{\mathrm{T}}_{\max}+1)$- and $(M^{\mathrm{E}}_{\max}+1)\times(M^{\mathrm{E}}_{\max}+1)$-matrices, and $\boldsymbol{\alpha}_{\mathrm{T}}$ and $\boldsymbol{\alpha}_{\mathrm{E}}$ are $((M^{\mathrm{T}}_{\max}+1)\times 1)$- and $((M^{\mathrm{E}}_{\max}+1)\times 1)$-vectors of parameters, respectively. However, to simply our model in Eq. (5.11), PRSS can be approximated subsequently by using a single multiplier of penalization:

$$PRSS \approx \left\| \widehat{\boldsymbol{X}} - \breve{\boldsymbol{X}} \right\|_2^2 + \phi \left\| \boldsymbol{L} \boldsymbol{\alpha} \right\|_2^2, \tag{5.12}$$

where $\breve{\boldsymbol{X}} = (\breve{\boldsymbol{X}}^T, \breve{\boldsymbol{E}}^T)^T$, $\widehat{\boldsymbol{X}} = (\widehat{\boldsymbol{X}}^T, \widehat{\boldsymbol{E}}^T)^T$ and $\boldsymbol{L} = (\boldsymbol{L}_{\mathrm{T}}, \boldsymbol{L}_{\mathrm{E}})$. Here, $M_{\max} = M^{\mathrm{T}}_{\max} + M^{\mathrm{E}}_{\max}$, and $\boldsymbol{\alpha} = (\boldsymbol{\alpha}^{\mathrm{T}}_{\mathrm{T}}, \boldsymbol{\alpha}^{\mathrm{T}}_{\mathrm{E}})^T$ is an $((M_{\max}+1)\times 1)$-vector of parameters to be estimated with the help of the data points. Consequently, for target-environment networks, we may represent our optimization problem in the following form (with reference to Eq. (5.8)):

$$\begin{aligned} \underset{w,\boldsymbol{\alpha}}{\text{minimize}} \quad & w, \\ \text{subject to} \quad & \left\| \widehat{\boldsymbol{X}} - \breve{\boldsymbol{X}} \right\|_2 \leq w, \\ & \left\| \boldsymbol{L} \boldsymbol{\alpha} \right\|_2 \leq \sqrt{K}, \end{aligned} \tag{5.13}$$

with some chosen parameter $K \geq 0$.

5.1.1 Polyhedral Uncertainty and Robust Counterpart for Regulatory Networks

To evaluate and solve the robustness problem, for target-environment networks, we suppose that the model uncertainty is represented by a family of matrices $\undertilde{\widetilde{X}} = \widetilde{X} + U_{\mathrm{T}}, \undertilde{\widetilde{E}} = \widetilde{E} + U_{\mathrm{E}}$ and vectors $\widehat{\widehat{X}} = \widehat{X} + v_{\mathrm{T}}, \widehat{\widehat{E}} = \widehat{E} + v_{\mathrm{E}}$, where $U = (U_{\mathrm{T}}, U_{\mathrm{E}}) \in U_1 := (U_1^{\mathrm{T}} \times U_1^{\mathrm{E}})$ and $v = (v_{\mathrm{T}}, v_{\mathrm{E}}) \in U_2 := (U_2^{\mathrm{T}} \times U_2^{\mathrm{E}})$ are unknown matrices and vectors but they are situated in bounded sets, respectively. These uncertainty matrices $U \in U_1$ and uncertainty vectors $v \in U_2$ are by

$$U = \begin{bmatrix} u_{0,1}^{\mathrm{T}} & u_{0,2}^{\mathrm{T}} & \cdots & u_{0,M_{\max}}^{\mathrm{T}} & u_{0,1}^{\mathrm{E}} & u_{0,2}^{\mathrm{E}} & \cdots & u_{0,M_{\max}}^{\mathrm{E}} \\ u_{1,1}^{\mathrm{T}} & u_{1,2}^{\mathrm{T}} & \cdots & u_{1,M_{\max}}^{\mathrm{T}} & u_{1,1}^{\mathrm{E}} & u_{1,2}^{\mathrm{E}} & \cdots & u_{1,M_{\max}}^{\mathrm{E}} \\ \vdots & \vdots & \ddots & \vdots & \vdots & \vdots & \ddots & \vdots \\ u_{N,1}^{\mathrm{T}} & u_{N,2}^{\mathrm{T}} & \cdots & u_{N,M_{\max}}^{\mathrm{T}} & u_{N,1}^{\mathrm{E}} & u_{N,2}^{\mathrm{E}} & \cdots & u_{N,M_{\max}}^{\mathrm{E}} \end{bmatrix}, \tag{5.14}$$

$$v = \left((v_0^{\mathrm{T}}\ v_1^{\mathrm{T}}\ \ldots\ v_N^{\mathrm{T}})^T, (v_0^{\mathrm{E}}\ v_1^{\mathrm{E}}\ \ldots\ v_N^{\mathrm{E}})^T \right). \tag{5.15}$$

Based on those underlying sets U_1 and U_2, the robust counterpart is determined as follows:

$$\underset{\alpha_{\mathrm{T}},\alpha_{\mathrm{E}}}{\text{minimize}} \max_{\substack{(\widetilde{W}_{\mathrm{T}},\widetilde{W}_{\mathrm{E}}) \in U_1, \\ (\hat{z}_{\mathrm{T}},\hat{z}_{\mathrm{E}}) \in U_2}} \left\| \hat{z}_{\mathrm{T}} - \widetilde{W}_{\mathrm{T}} \alpha_{\mathrm{T}} \right\|_2^2 + \left\| \hat{z}_{\mathrm{E}} - \widetilde{W}_{\mathrm{E}} \alpha_{\mathrm{E}} \right\|_2^2 + \phi_{\mathrm{T}} \left\| L_{\mathrm{T}} \alpha_{\mathrm{T}} \right\|_2^2 + \phi_{\mathrm{E}} \left\| L_{\mathrm{E}} \alpha_{\mathrm{E}} \right\|_2^2. \tag{5.16}$$

Namely, U_1 is a polytope with $2^{(N+1)M_{\max}}$ vertices $\widetilde{W}^1, \widetilde{W}^2, \ldots, \widetilde{W}^{2^{(N+1)M_{\max}}}$ and represented as

$$U_1 = \left\{ \sum_{\kappa=1}^{2^{(N+1)M_{\max}}} \delta_\kappa \widetilde{W}^\kappa \mid \delta_\kappa \geq 0\ (\kappa \in \{1, 2, \ldots, 2^{(N+1)M_{\max}}\}), \sum_{\kappa=1}^{2^{(N+1)M_{\max}}} \delta_\kappa = 1 \right\}, \tag{5.17}$$

where $U_1 = \operatorname{conv}\left\{ \widetilde{W}, \widetilde{W}^2, \ldots, \widetilde{W}^{2^{(N+1)M_{\max}}} \right\}$. Furthermore, U_2 is a polytope with 2^{N+1} vertices $\hat{z}^1, \hat{z}^2, \ldots, \hat{z}^{2^{(N+1)}}$ and it can be expressed as

$$U_2 = \left\{ \sum_{\mu=1}^{2^{(N+1)}} \varphi_\mu \hat{z}^\mu \mid \varphi_\mu \geq 0\ (\mu \in \{1, 2, \ldots, 2^{(N+1)}\}), \sum_{\mu=1}^{2^{(N+1)}} \varphi_\mu = 1 \right\}, \tag{5.18}$$

i.e., $U_2 = \operatorname{conv}\left\{ \hat{z}^1, \hat{z}^2, \ldots, \hat{z}^{2^{(N+1)}} \right\}$. The uncertainty sets U_1 and U_2 have the form of polytopes and they can be presented as a convex combination of vertices $\widetilde{W}^\kappa$ ($\kappa =$

$1, 2, \ldots, 2^{(N+1)M_{\max}}$) and $\widetilde{z}^{\mu}$ ($\mu = 1, 2, \ldots, 2^{(N+1)}$), respectively. Now, the entries of $\widetilde{\boldsymbol{W}}$ and $\hat{z}$ may be thought to have become intervals, in fact, our CIs. Then, the matrix $\widetilde{\boldsymbol{W}}$ and vector $\hat{z}$ with uncertainty are lying in the Cartesian products of intervals; those are parallelpipes (for visualization, cf. Sect. 3.1.3).

5.1.2 Robust Conic Quadratic Programming with Polyhedral Uncertainty

When *polyhedral* uncertainty is employed by uncertainty sets U_1 and U_2, for our RCMARS model on target-environment networks, the robust CQP program is represented in the following manner:

$$
\begin{aligned}
&\underset{w_{\mathrm{T}}, w_{\mathrm{E}}, \boldsymbol{\alpha}_{\mathrm{T}}, \boldsymbol{\alpha}_{\mathrm{E}}}{\text{minimize}} \; w_{\mathrm{T}} + w_{\mathrm{E}}, \\
&\text{subject to} \left\| \hat{z}_{\mathrm{T}} - \widetilde{\boldsymbol{W}}_{\mathrm{T}} \boldsymbol{\alpha}_{\mathrm{T}} \right\|_2 \le w_{\mathrm{T}}, \\
&\qquad \left\| \hat{z}_{\mathrm{E}} - \widetilde{\boldsymbol{W}}_{\mathrm{E}} \boldsymbol{\alpha}_{\mathrm{E}} \right\|_2 \le w_{\mathrm{E}}, \ \forall \ \underbrace{(\widetilde{\boldsymbol{W}}^{\mathrm{T}}, \widetilde{\boldsymbol{W}}^{\mathrm{E}})}_{=\sum_{\kappa=1}^{2^{(N+1)M_{\max}}} \delta_{\kappa} \widetilde{\boldsymbol{W}}^{\kappa}} \in U_1, \quad \underbrace{(\hat{z}^{\mathrm{T}}, \hat{z}^{\mathrm{E}})}_{=\sum_{\mu=1}^{2^{N+1}} \varphi_{\mu} \hat{z}^{\mu}} \in U_2, \\
&\qquad \| \boldsymbol{L} \boldsymbol{\alpha}_{\mathrm{T}} \|_2 \le \sqrt{K_{\mathrm{T}}}, \\
&\qquad \| \boldsymbol{L} \boldsymbol{\alpha}_{\mathrm{E}} \|_2 \le \sqrt{K_{\mathrm{E}}}.
\end{aligned}
\tag{5.19}
$$

Since U_1 and U_2 are polytopes, described by their vertices as

$$
U_1 = \operatorname{conv}\left\{ \widetilde{\boldsymbol{W}}^1, \widetilde{\boldsymbol{W}}^2, \ldots, \widetilde{\boldsymbol{W}}^{2^{(N+1)M_{\max}}} \right\}, \ U_2 = \operatorname{conv}\left\{ \hat{z}^1, \hat{z}^2, \ldots, \hat{z}^{2^{N+1}} \right\},
$$

then our robust CQP can be equivalently expressed as a *standard* CQP [35] with the subsequent form:

$$
\begin{aligned}
&\underset{w_{\mathrm{T}}, w_{\mathrm{E}}, \boldsymbol{\alpha}_{\mathrm{T}}, \boldsymbol{\alpha}_{\mathrm{E}}}{\text{minimize}} \; w_{\mathrm{T}} + w_{\mathrm{E}} \\
&\text{subject to} \left\| \hat{z}_{\mathrm{T}}^{\mu} - \widetilde{\boldsymbol{W}}_{\mathrm{T}}^{\kappa_1} \boldsymbol{\alpha}_{\mathrm{T}} \right\|_2 \le w_{\mathrm{T}} \ (\mu = 1, 2, \ldots, 2^{N+1}; \kappa_1 = 1, 2, \ldots, 2^{(N+1)M_{\max}^{\mathrm{T}}}), \\
&\qquad \left\| \hat{z}_{\mathrm{E}}^{\mu} - \widetilde{\boldsymbol{W}}_{\mathrm{E}}^{\kappa_2} \boldsymbol{\alpha}_{\mathrm{E}} \right\|_2 \le w_{\mathrm{E}} \ (\mu = 1, 2, \ldots, 2^{N+1}; \kappa_2 = 1, 2, \ldots, 2^{(N+1)M_{\max}^{\mathrm{E}}}), \\
&\qquad \| \boldsymbol{L} \boldsymbol{\alpha}_{\mathrm{T}} \|_2 \le \sqrt{K_{\mathrm{T}}}, \\
&\qquad \| \boldsymbol{L} \boldsymbol{\alpha}_{\mathrm{E}} \|_2 \le \sqrt{K_{\mathrm{E}}}.
\end{aligned}
\tag{5.20}
$$

Then, to facilitate our representation in Eq. (5.20), this problem can be rewritten as

$$\begin{aligned} &\underset{w,\alpha}{\text{minimize}}\ w, \\ &\text{subject to} \left\| \hat{z}^{\mu} - \widetilde{\boldsymbol{W}}^{\kappa}\boldsymbol{\alpha} \right\|_2 \leq w, \\ &\qquad\qquad \|\boldsymbol{L}\boldsymbol{\alpha}\|_2 \leq \sqrt{K}, \end{aligned} \tag{5.21}$$

where $\mu = 1,\ldots,2^{N+1}; \kappa = (\kappa_1,\kappa_2) \in \prod_{c=\mathrm{T}}^{\mathrm{E}} \{1,\ldots,2^{(N+1)M_{\max}^{c}}\}$. Hence, $\widetilde{\boldsymbol{W}}^{\kappa} = \begin{pmatrix} \widetilde{\boldsymbol{W}}_{\mathrm{T}}^{\kappa_1} & \boldsymbol{0} \\ \boldsymbol{0} & \widetilde{\boldsymbol{W}}_{\mathrm{E}}^{\kappa_2} \end{pmatrix}$, $\hat{z}^{\mu} = (\hat{z}_{\mathrm{T}}^{\mu T}, \hat{z}_{\mathrm{E}}^{\mu T})^T$, $\boldsymbol{\alpha} = (\boldsymbol{\alpha}_{\mathrm{T}}{}^T, \boldsymbol{\alpha}_{\mathrm{E}}{}^T)^T$ and $\boldsymbol{L} = (\boldsymbol{L}_{\mathrm{T}}, \boldsymbol{L}_{\mathrm{E}})$. Here, $\boldsymbol{0}$, which can have different formats, is used a dummy variable of $\boldsymbol{0}$- matrices to simplify notation.

5.2 Numerical Experience

5.2.1 Developing RCMARS Models for Regulatory Networks

For an implementation example of the RCMARS algorithm within our dynamical model application, we refer to an artificial dataset that has 2 targets and 2 environmental variables. So, we have 4 predictor variables $(\widetilde{x}_1, \widetilde{x}_2, \widetilde{e}_1, \widetilde{e}_2)$ with 25 measurement values for each of them. Based on that, the maximum number of BFs, $M_{\max}$, and the highest degree of interaction are determined for each targets and environmental items, and the largest models are constructed in the forward MARS algorithm by its software, *Salford MARS* [83] (cf. Sect. 2.3.3). To prevent from nondifferentiability in our optimization program, we choose the knot values different from the data points, but these values to be very much nearby to the corresponding input data. Hence, for both targets and environmental factors, the numbers M_{T} of BFs are 11, 10, 8, and 11, respectively, and the largest models of RCMARS become

$$\begin{aligned} \hat{x}_1 = \alpha_0 &+ \alpha_1 \max\{0, \widetilde{e}_2 + 2.113\} + \alpha_2 \max\{0, \widetilde{e}_1 + 2.106\} + \alpha_3 \max\{0, \widetilde{x}_1 - 2.337\} \\ &+ \alpha_4 \max\{0, \widetilde{x}_2 - 0.058\} + \alpha_5 \max\{0, 0.058 - \widetilde{x}_2\} + \alpha_6 \max\{0, \widetilde{x}_1 + 0.295\} \\ &+ \alpha_7 \max\{0, -0.295 - \widetilde{x}_1\} + \alpha_8 \max\{0, \widetilde{e}_2 + 0.079\} + \alpha_9 \max\{0, -0.079 - \widetilde{e}_1\} \\ &+ \alpha_{10} \max\{0, \widetilde{e}_1 + 0.195\} + \alpha_{11} \max\{0, -0.195 - \widetilde{e}_1\}, \end{aligned}$$

$$\begin{aligned} \hat{x}_2 = \alpha_0 &- \alpha_1 \max\{0, \widetilde{e}_1 + 2.106\} + \alpha_2 \max\{0, \widetilde{e}_2 - 0.392\} + \alpha_3 \max\{0, 0.392 - \widetilde{e}_2\} \\ &+ \alpha_4 \max\{0, \widetilde{x}_2 + 1.838\} + \alpha_5 \max\{0, \widetilde{x}_1 + 0.295\} + \alpha_6 \max\{0, -0.295 - \widetilde{x}_1\} \end{aligned}$$

$$+ \alpha_7 \max\{0, \widetilde{x}_2 - 0.058\} + \alpha_8 \max\{0, 0.058 - \widetilde{x}_2\} + \alpha_9 \max\{0, \widetilde{x}_2 + 0.347\}$$
$$+ \alpha_{10} \max\{0, -0.347 - \widetilde{x}_2\},$$

$$\begin{aligned}\hat{e}_1 = {} & \alpha_0 + \alpha_1 \max\{0, \widetilde{x}_1 + 2.337\} + \alpha_2 \max\{0, \widetilde{e}_1 + 2.195\} + \alpha_3 \max\{0, -0.195 - \widetilde{e}_1\} \\ & + \alpha_4 \max\{0, \widetilde{x}_2 + 1.838\} + \alpha_5 \max\{0, \widetilde{e}_2 + 0.010\} \cdot \max\{0, \widetilde{x}_1 + 2.337\} \\ & + \alpha_6 \max\{0, -0.010 - \widetilde{e}_2\} \cdot \max\{0, \widetilde{x}_1 + 2.337\} \\ & + \alpha_7 \max\{0, \widetilde{x}_1 + 0.295\} \cdot \max\{0, \widetilde{x}_2 + 1.838\} \\ & + \alpha_8 \max\{0, -0.295 - \widetilde{x}_1\} \cdot \max\{0, \widetilde{x}_2 + 1.838\},\end{aligned}$$

$$\begin{aligned}\hat{e}_2 = {} & \alpha_0 + \alpha_1 \max\{0, \widetilde{e}_2 + 2.113\} + \alpha_2 \max\{0, \widetilde{e}_1 - 0.450\} + \alpha_3 \max\{0, 0.450 - \widetilde{e}_1\} \\ & + \alpha_4 \max\{0, \widetilde{x}_1 + 0.295\} + \alpha_5 \max\{0, -0.295 - \widetilde{x}_1\} + \alpha_6 \max\{0, \widetilde{x}_2 - 0.058\} \\ & + \alpha_7 \max\{0, 0.058 - \widetilde{x}_2\} + \alpha_8 \max\{0, \widetilde{e}_1 + 0.195\} + \alpha_9 \max\{0, -0.195 - \widetilde{e}_1\} \\ & + \alpha_{10} \max\{0, \widetilde{x}_2 + 0.347\} + \alpha_{11} \max\{0, -0.347 - \widetilde{x}_2\}.\end{aligned}$$

As our next step, for the second part of our optimization model in Eq. (5.12) the matrices $\boldsymbol{L}$ are obtained, related to all targets and environmental factors, respectively. To introduce the robust optimization approach into the RCMARS model, by applying Eq. (5.7), uncertainties are calculated for all input and output values which are represented by CIs, and these uncertainty values evaluated are inserted into the real input data $\widetilde{\boldsymbol{x}}_k$ and $\widetilde{\boldsymbol{e}}_k$ in each dimension, and into the output data $\hat{\boldsymbol{x}}_k$ and $\hat{\boldsymbol{e}}_k$ $(k = 0, 1, \ldots, 24)$. Therefore, for both targets and environmental items, the uncertainty matrices and vectors based on polyhedral uncertainty sets are constructed by using Eqs. (5.17)–(5.18). Indeed, we have a tradeoff here between tractability and robustification, because , the uncertainty matrices of the input data have huge dimensions, and we do not possess enough computer capacity to solve our problem with respect to these uncertainty matrices. To cope with this difficulty, for all targets and environmental items, we formulate the minimization of PRSS as a CQP problem in Eq. (5.21) for all data values by following a *combinatorial approach* that we call *weak robustification* (cf. Remark 2).

As a result, we obtain 25 different *weak RCMARS (WRCMARS)* models for both targets and environmental items. These 100 ($= 25 \cdot 4$) sub-models are solved independently by running the program code of RCMARS algorithm written in MATLAB and using MOSEK software [87] for CQP problem, and we receive the w value for each of our auxiliary problems. Eventually, as an expression of our *worst*-case approach, we chose the solution that has the *maximum* w value, in terms of all targets and environmental factors. For our RCMARS involvement, Table 5.1 displays the optimal parameters of targets and environmental factors found.

Table 5.1 For targets and environmental factors, predicted parameter values by RCMARS algorithm

	α_0	α_1	α_2	α_3	α_4	α_5
$\tilde{x}_1$	−0.247	0.111	−0.326	0.269	0.191	−0.050
$\tilde{x}_2$	0.711	−0.448	−0.924	0.366	0.130	−0.097
$\tilde{e}_1$	−2.258	0.782	0.549	−0.392	0.147	0.000
$\tilde{e}_2$	−1.708	0.616	−0.077	0.434	0.522	−0.230
	α_6	α_7	α_8	α_9	α_{10}	α_{11}
$\tilde{x}_1$	0.382	−0.314	0.201	−0.217	−0.444	0.215
$\tilde{x}_2$	0.000	0.104	0.030	0.033	−0.112	
$\tilde{e}_1$	−0.056	0.000	0.066			
$\tilde{e}_2$	−0.080	−0.085	−0.292	0.384	0.013	−0.015

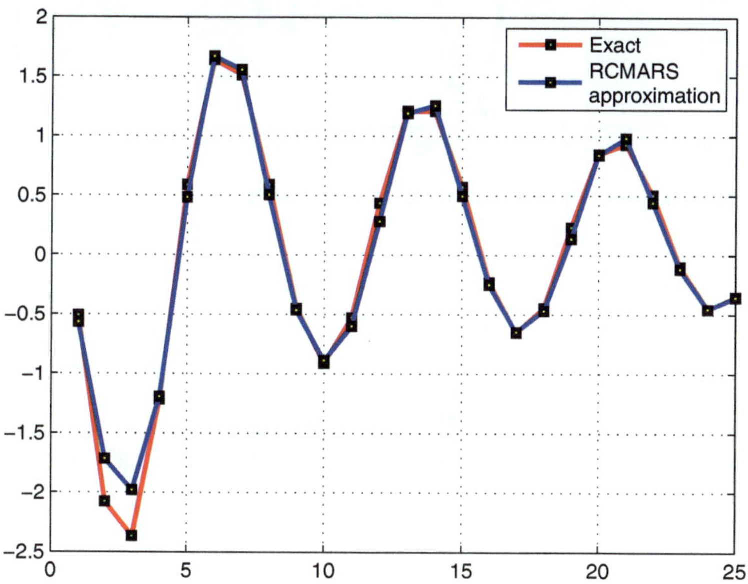

Fig. 5.3 True and predicted expression values of the first target

5.2.2 Results

The prediction results for targets and environmental factors can be seen in Figs. 5.3, 5.4, 5.5, 5.6; here, the "red line" presents exact values, and the "blue line" indicates the predicted values by RCMARS model.

From Figs. 5.3–5.6, with the exact expression data of targets and environmental factors, we may deduce that the predicted values of RCMARS model match very well. This implies that with RCMARS our **new robust** regression model for

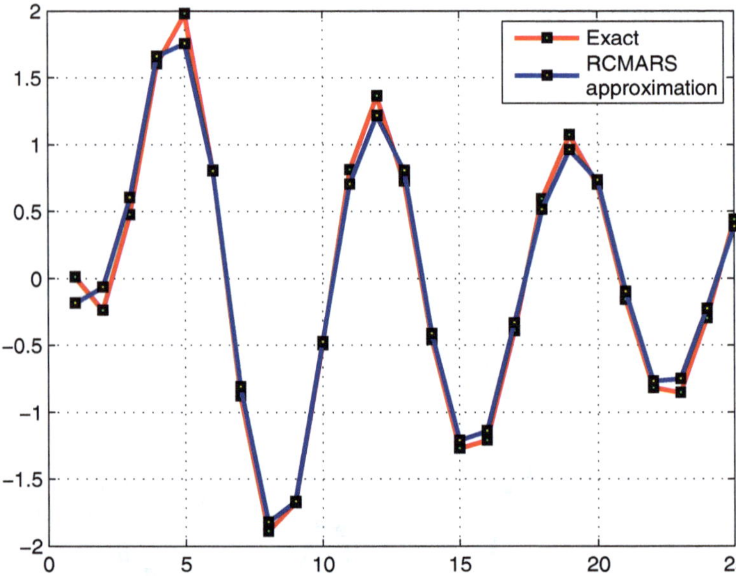

Fig. 5.4 True and predicted expression values of the second target

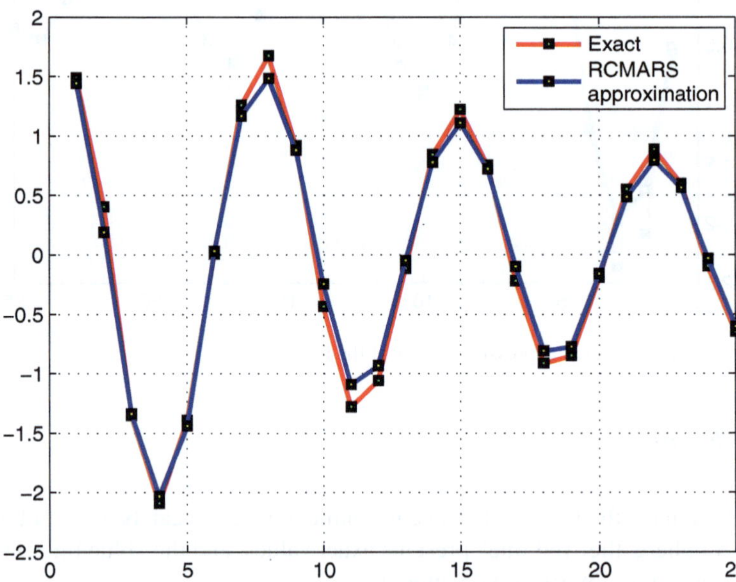

Fig. 5.5 True and predicted expression values of the first environmental item

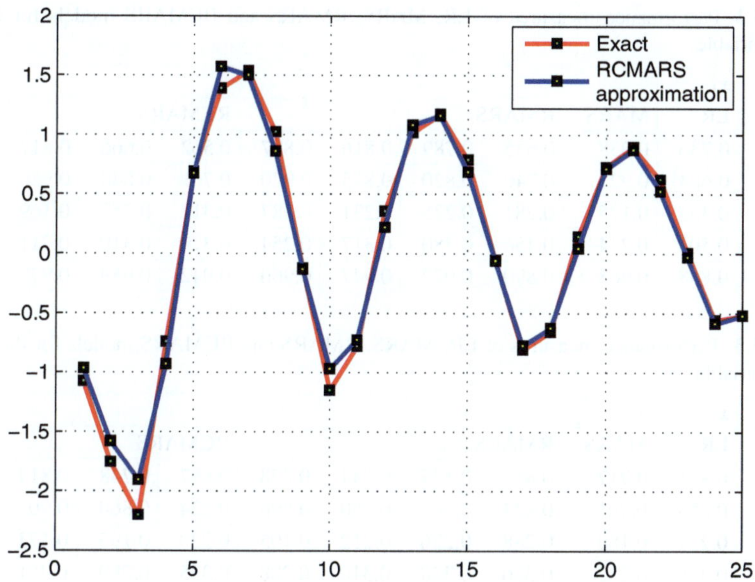

Fig. 5.6 True and predicted expression values of the second environmental item

regulatory systems can predict the trend of the target-environment interaction very successfully.

5.2.3 *Simulation Study and Comparison*

In previous subsections, we presented and scientifically analyzed R(C)MARS and two-model regulatory systems under polyhedral uncertainty. What is more, using an artificial data set, we introduced RCMARS models for 2 targets and 2 environmental factors as a numerical experience and represented the results obtaining figures for each target and each environmental items. In this subsection, to demonstrate the performance of RCMARS for a regulatory system based on replicated datasets and compare this method with other related methods. We construct different LR, MARS and RCMARS models through 5 different simulated datasets for each target and each environmental item as we defined in Sect. 5.2. In the study [106], our basic performance measure to calculate the precision of the models is *estimation variance (EV)*. According to our main aim, we evaluate EVs for LR, MARS and RCMARS models. Also, to compare the results concerning the accuracy of LR, MARS, and RCMARS, these models are evaluated based on some accuracy measures such as R^2_{adj}, *AAE*, *RMSE*, and *r*. The explanations, interpretations and formulas of these measures are presented in Table D.1. When developing RCMARS models, a sensitivity study is conducted to determine the most appropriate confidence limits

Table 5.2 Performance measures of LR, MARS, RMARS and RCMARS models for the first target variable

	x_1									
	LR	MARS	RMARS				RCMARS			
EV	0.736	0.938	0.635	0.789	0.816	0.887	0.562	0.666	0.817	0.852
R^2_{adj}	0.684	0.925	0.740	0.820	0.875	0.920	0.761	0.840	0.926	0.940
AAE	0.333	0.175	0.281	0.275	0.231	0.187	0.317	0.257	0.169	0.151
$RMSE$	0.503	0.244	0.456	0.380	0.317	0.254	0.379	0.310	0.211	0.190
r	0.858	0.968	0.890	0.922	0.947	0.966	0.942	0.959	0.979	0.983

Table 5.3 Performance measures of LR, MARS, RMARS and RCMARS models for the second target variable

	x_2									
	LR	MARS	RMARS				RCMARS			
EV	0.871	0.917	0.813	0.835	0.844	0.878	0.697	0.748	0.819	0.866
R^2_{adj}	0.859	0.901	0.824	0.865	0.880	0.896	0.824	0.860	0.902	0.925
AAE	0.231	0.189	0.288	0.226	0.212	0.195	0.220	0.195	0.173	0.161
$RMSE$	0.336	0.282	0.376	0.328	0.310	0.288	0.335	0.299	0.251	0.219
r	0.939	0.958	0.924	0.943	0.949	0.856	0.946	0.957	0.969	0.976

Table 5.4 Performance measures of LR, MARS, RMARS and RCMARS models for the first environmental variable

	e_1									
	LR	MARS	RMARS				RCMARS			
EV	0.839	0.912	0.665	0.682	0.775	0.888	0.590	0.664	0.888	0.907
R^2_{adj}	0.809	0.894	0.810	0.842	0.877	0.893	0.780	0.810	0.895	0.901
AAE	0.268	0.243	0.328	0.308	0.265	0.247	0.268	0.241	0.201	0.196
$RMSE$	0.391	0.291	0.390	0.356	0.314	0.292	0.375	0.349	0.260	0.252
r	0.917	0.955	0.925	0.940	0.951	0.955	0.939	0.943	0.965	0.966

on both the input and output data. For this aim, different uncertainty matrices, $\boldsymbol{U}$, for the input data and different uncertainty vectors, $\boldsymbol{v}$, for the output data in Eqs. (5.14)–(5.15) are obtained by using different intervals and R(C)MARS results are represented based on *four different uncertainty scenarios*.

According to all the aforementioned computations and comparisons, our R(C)MARS method proves to be very competitive with the other methods. We are able to achieve a variance reduction which is very important in practice, and an additional advantage, especially, when comparing with our predecessor method of MARS. On the other hand, as it is deduced in Tables 5.2, 5.3, 5.4, 5.5 and those performance criteria, in general, RCMARS produced more accurate models with smaller variances than LR and MARS and RMARS with respect to precision and accuracy. Consequently, R(C)MARS can provide us very good predictions for the dynamics of the target-environment interaction based on the expression values of

Table 5.5 Performance measures of LR, MARS, RMARS and RCMARS models for the second environmental variable

	e_2									
	LR	MARS	RMARS				RCMARS			
EV	0.848	0.860	0.665	0.727	0.748	0.814	0.620	0.667	0.704	0.834
R^2_{adj}	0.818	0.840	0.784	0.818	0.825	0.838	0.785	0.819	0.842	0.910
AAE	0.275	0.285	0.339	0.310	0.302	0.287	0.241	0.221	0.211	0.166
RMSE	0.382	0.367	0.426	0.391	0.383	0.369	0.382	0.351	0.328	0.248
r	0.921	0.927	0.905	0.919	0.922	0.927	0.932	0.942	0.949	0.969

both all targets and all environmental factors. Therefore, we indicate that RCMARS can perform better than LR, MARS for all target and environmental items with respect to any of our measures, as validated through simulated datasets.

Here, the performance of RMARS and RCMARS are compared by using only one simulated dataset and the results of RMARS and RCMARS models with LR and MARS are demonstrated in Tables 5.2–5.5 based on *four different uncertainty scenarios*. Indeed, these results and the results which we demonstrated in our previous chapter deduce that CMARS performs better than MARS, and thus RCMARS performs better than RMARS for all target and environmental items with respect to all measures validated through simulated datasets. Therefore, we continue comparing the performance of RCMARS with LR and MARS through the remaining 4 different simulated datasets and represent the results of LR, MARS and RCMARS models in Tables B.1 and B.2.

Chapter 6
Real-World Application with Our Robust Tools

6.1 A Real-World Application of RCMARS in the Financial Sector

6.1.1 Introduction

One of the fundamental concepts in finance theory is optimization, and the financial decision making for a rational agent is essentially a question of achieving an optimal trade-off between risk and return. In this way, robustification is starting to draw more attention in finance; in particular, some studies report promising results using robust statistical techniques in financial markets. In the study [101], we used data from Istanbul Stock Exchange like ISE 100 index, ISE transaction number and so on, from Turkish economy like TUFE and TEFE indexes, and also data of the Fed Funds Interest Rate and VIX Index which have been obtained from the US market, because of their strong effect on the economy of Turkey. ISE 100 index has been taken as the dependent variable, and others as the independent variables. We put a correlation threshold in order to limit the unnecessary and meaningless calculations and eliminated several variables which do not satisfy this requirement. Afterwards, we applied RCMARS to the remaining independent variables.

6.1.2 Data Description

We selected our time-series data for the empirical part from the website of Central Bank of the Republic of Turkey [25]. The data contain the economic indicators which are the most commonly used ones for the interpretation of an economic situation. Monthly data have been preferred in order to have more definite and stationary results, relative to daily or weekly data. If we could not find the monthly

A. Özmen, *Robust Optimization of Spline Models and Complex Regulatory Networks*, Contributions to Management Science,
DOI 10.1007/978-3-319-30800-5_6

data, we used daily data and converted them to monthly data by taking averages, or for some of them the last data of the month were taken as the data of the month, like Net Foreign Exchange Reserves and International Gold Reserves. *ISE 100 stock index* is the dependent variable in our dataset. We used this *index*, because it is a statistical measure of change in an economy or a securities market. For financial markets, an index is an imaginary portfolio of securities representing a particular market or a portion of it. It has its own calculation methodology and is usually expressed in terms of a change from a base value. Thus, the percentage change is more important than the actual numerical value.

The independent variables are *ISE Transaction Number* (the number of transaction during a defined time period, in our case during the month), *ISE Trading Volume* (the number of shares or contracts of a security traded during a defined time period, again for a month), *Capacity Usage Ratio* (the ratio of the production capacity of the economy to the maximum capacity of economy), *Euro and Dollar Exchange Rate*, *Net Foreign Exchange Reserves and International Gold Reserves*, *Gold Price*, *Credit Volume*, *Price Indexes* like *Wholesale Price Index* (*WPI*) and *Consumer Price Index* (*CPI*) (in Turkey: TEFE and TUFE, respectively). WPI is the price of a representative basket of wholesale goods, while a CPI measures changes in the price level of consumer goods and services purchased by households. Two indicators from the USA are taken to our analysis: *Fed Funds Interest Rate* and *VIX Index* (a measure of the market's expectation of stock market volatility over the next 30 day period), because of the strong effect of the USA on the economy of Turkey and the world. We use ISE 100 Stock Market index as a dependent variable. This is the successor of the *Composite Index*, which was introduced in 1986 including the stocks of 40 companies and was in time limited to the stocks of 100 companies. It consists of 100 stocks, which have been selected among the stocks of companies listed on the National Market, and the stocks of real estate investment trusts and venture capital investment trusts, listed on the Corporate Products Market, and it covers ISE 30 and ISE 50 stocks.

The data cover the time horizon between January 1999 and December 2009. Some of the series do not contain the data of December 2009; therefore, the absent values are calculated in Excel using interpolation. We also checked the correlation among these series, in order to prevent from unnecessary and meaningless calculations. We assumed a correlation threshold of 0.90 to decide about the strength of correlation. The most correlated factors are ISE Trading Volume, International Gold Reserves, Net Foreign Exchange Reserves and WPI (TEFE). For example, there is a correlation of 0.94 between ISE Transaction Number and ISE Trading Volume. So, ISE Transaction Number is taken out from the list. Eventually, our dataset consists of ISE Trading Volume, Capacity Usage Ratio, Euro and Dollar Exchange Rates, Credit Volume, Gold Price, WPI (TEFE), Fed Funds Interest Rate and VIX Index.

6.1.3 Obtaining Large Model from MARS Program

For the implementation of our RCMARS algorithm developed, we used a dataset from the financial market and, eliminating some of the predictor variables which have the correlation. At the end we have 8 predictor input variables:

X_1 : ISE Trading Volume, X_2 : Capacity Usage Ratio,

X_3 : Euro Exchange Rate, X_4 : Credit Volume,

X_5 : Dollar Exchange Rate, X_6 : Price Index (TEFE),

X_7 : Federal Funds Interest Rate, X_8 : VIX Index,

with 76 observations. However, we do not have enough computer capacity to solve our problem in Eq. (3.19) that is given as a *tradeoff* between tractability and robustification. Therefore we divide our dataset into two subsets, each of which has 38 observations. Firstly, we validate our assumption that the input variables and the output variable are distributed normally, using *bootstrapping method* [34] from statistics. In order to implement RCMARS algorithm, first, the MARS models are constructed for each subset by using the Salford MARS version 3 [83] and, then, the maximum number of BFs ($M_{\max}$) and the highest degree of interactions are determined by trial and error. In first part of our dataset, $M_{\max}$ is assigned to be 12, and the highest degree of interaction is assigned to be 3. Then, the largest models for the first part and the second part of the dataset are constructed in the forward MARS algorithm by its software.

To prevent from nondifferentiability in our optimization problem, we choose the knot values different from data points. However, these values are very much nearby to the corresponding input data. Then, the BFs for the first part of the dataset can be introduced into the largest model subsequent way[1]:

$$
\begin{aligned}
\hat{y} &= \alpha_0 + \sum_{m=1}^{M} \alpha_m \vartheta_m(\boldsymbol{x}) \\
&= \alpha_0 + \alpha_1 \vartheta_1(\boldsymbol{x}) + \alpha_2 \vartheta_2(\boldsymbol{x}) + \alpha_3 \vartheta_3(\boldsymbol{x}) + \alpha_4 \vartheta_4(\boldsymbol{x}) + \alpha_5 \vartheta_5(\boldsymbol{x}) + \alpha_6 \vartheta_6(\boldsymbol{x}) \\
&\quad + \alpha_7 \vartheta_7(\boldsymbol{x}) + \alpha_8 \vartheta_8(\boldsymbol{x}) + \alpha_9 \vartheta_9(\boldsymbol{x}) + \alpha_{10} \vartheta_{10}(\boldsymbol{x}) + \alpha_{11} \vartheta_{11}(\boldsymbol{x}) + \alpha_{12} \vartheta_{12}(\boldsymbol{x}) \\
&= \alpha_0 + \alpha_1 \max\{0, x_8 - 0.365\} + \alpha_2 \max\{0, 0.365 - x_8\} \\
&\quad + \alpha_3 \max\{0, x_1 + 0.567\} + \alpha_4 \max\{0, -0.567 - x_1\} \\
&\quad + \alpha_5 \max\{0, x_2 + 0.542\} + \alpha_6 \max\{0, -0.542 - x_2\}
\end{aligned}
$$

[1]For the ease of representation, here and subsequently, we suppress the index m of the subvectors $\boldsymbol{x}^m$ and just write $\boldsymbol{x}$.

$$
\begin{aligned}
&+ \alpha_7 \max\{0, x_4 + 2.187\} \cdot \max\{0, -0.542 - x_2\} \\
&+ \alpha_8 \max\{0, x_4 + 0.098\} \cdot \max\{0, 0.365 - x_8\} \\
&+ \alpha_9 \max\{0, -0.098 - x_4\} \cdot \max\{0, 0.365 - x_8\} \\
&+ \alpha_{10} \max\{0, x_7 + 2.216\} \cdot \max\{0, x_1 + 0.567\} \\
&+ \alpha_{11} \max\{0, x_6 - 0.542\} \cdot \max\{0, x_7 + 2.216\} \cdot \max\{0, x_1 + 0.567\} \\
&+ \alpha_{12} \max\{0, 0.542 - x_8\} \cdot \max\{0, x_7 + 2.216\} \cdot \max\{0, x_1 + 0.567\}.
\end{aligned}
$$

Likewise, the BFs for the second part of the dataset become inserted in the largest model in the following manner:

$$
\begin{aligned}
\hat{y} &= \alpha_0 + \sum_{m=1}^{M} \alpha_m \vartheta_m(\boldsymbol{x}) \\
&= \alpha_0 + \alpha_1 \vartheta_1(\boldsymbol{x}) + \alpha_2 \vartheta_2(\boldsymbol{x}) + \alpha_3 \vartheta_3(\boldsymbol{x}) + \alpha_4 \vartheta_4(\boldsymbol{x}) + \alpha_5 \vartheta_5(\boldsymbol{x}) + \alpha_6 \vartheta_6(\boldsymbol{x}) \\
&\quad + \alpha_7 \vartheta_7(\boldsymbol{x}) + \alpha_8 \vartheta_8(\boldsymbol{x}) + \alpha_9 \vartheta_9(\boldsymbol{x}) + \alpha_{10} \vartheta_{10}(\boldsymbol{x}) + \alpha_{11} \vartheta_{11}(\boldsymbol{x}) + \alpha_{12} \vartheta_{12}(\boldsymbol{x}) \\
&= \alpha_0 + \alpha_1 \max\{0, x_4 - 0.575\} + \alpha_2 \max\{0, 0.575 - x_3\} \\
&\quad + \alpha_5 \max\{0, x_1 - 0.019\} \cdot \max\{0, 0.275 - x_3\} \\
&\quad + \alpha_6 \max\{0, 0.019 - x_1\} \cdot \max\{0, 0.275 - x_3\} \\
&\quad + \alpha_7 \max\{0, x_1 + 2.172\} \cdot \max\{0, x_4 - 0.575\} \\
&\quad + \alpha_8 \max\{0, x_7 + 0.583\} \cdot \max\{0, 0.575 - x_4\} \\
&\quad + \alpha_9 \max\{0, x_5 + 0.309\} \cdot \max\{0, x_7 + 2.583\} \cdot \max\{0, 0.575 - x_4\} \\
&\quad + \alpha_{10} \max\{0, -0.309 - x_5\} \cdot \max\{0, x_7 + 2.583\} \cdot \max\{0, 0.575 - x_4\} \\
&\quad + \alpha_{11} \max\{0, x_2 + 0.499\} \cdot \max\{0, 0.575 - x_4\} \\
&\quad + \alpha_{12} \max\{0, -0.499 - x_2\} \cdot \max\{0, 0.575 - x_4\}.
\end{aligned}
$$

6.1.4 Bootstraping

In general, bootstrapping is used for statistical inference on the basic idea of building a sampling distribution for a statistic by resampling from the data at hand. It is also used to anticipate important characteristics of the population. Frequently mentioned comment about bootstrap is the following: '*The population is to the sample as the sample is to the bootstrap samples*'. The bootstrap provides correct statistical inference and is useful in driving accurate standard errors, confidence intervals and hypothesis tests for most statistics. It has also applicability in stratification, clustering by resampling from the sample data in the same wise as the original sample is selected from the population [34, 44].

6.1.5 Evaluating Accuracy and Complexity of PRSS Form

For this numeric example, we approximate the PRSS formula as follows:

$$PRSS \approx \overbrace{\left\| y - \vartheta(b)\alpha \right\|_2^2}^{=Accuracy} + \overbrace{\phi \left\| L\alpha \right\|_2^2}^{=Complexity} . \tag{6.1}$$

Herein, the first part of the TR term, which is the right-hand side, and that of the PRSS function, are equal to each other, whereas, their second parts are equal approximately. Subsequently, all those parts are stated:

Accuracy:

$$\left\| y - \vartheta(b)\alpha \right\|_2^2 = (y - \alpha^T \vartheta(b))^T (y - \alpha^T \vartheta(b)) = \sum_{k=1}^{N} (y_k - \alpha^T \vartheta(b_k))^2 =: (*), \tag{6.2}$$

Complexity:

$$\phi \left\| L\alpha \right\|_2^2 \approx \sum_{m=1}^{12} \phi_m \sum_{\substack{|\theta|=1 \\ \theta^T=(\theta_1,\theta_2)}}^{2} \sum_{\substack{r<s \\ r,s\in V(m)}} \int_{Q^m} \alpha_m^2 [D_{rs}^{\theta} \vartheta_m(t^m)]^2 dt^m =: (**), \tag{6.3}$$

where, indeed, $PRSS := (*) + (**)$ and $\phi = \phi_m$ $(m = 1, 2, \ldots, 12)$. Having discretized all the multi-dimensional integrals in the ***complexity*** part, they jointly turn into the form of Eq. (3.17) and, finally, the discretized form is indicated by $\boldsymbol{L}$. As a result, the matrix $\boldsymbol{L}$ becomes a diagonal matrix and the first column elements of $\boldsymbol{L}$ are all zero. The diagonal elements of this matrix, L_m $(m = 1, 2, \ldots, 12)$, are given below for the first part of our dataset:

$$\boldsymbol{L} = \begin{bmatrix} 0 & 0 & \ldots & 0 \\ 0 & 1.30 & \ldots & 0 \\ \vdots & \vdots & \ddots & \vdots \\ 0 & 0 & \ldots & 0.29 \end{bmatrix}.$$

For the second part of our dataset, the diagonal elements of $\boldsymbol{L}, L_m$ $(m = 1, 2, \ldots, 12)$ are comprised as follows:

$$\boldsymbol{L} = \begin{bmatrix} 0 & 0 & \ldots & 0 \\ 0 & 1.18 & \ldots & 0 \\ \vdots & \vdots & \ddots & \vdots \\ 0 & 0 & \ldots & 2.34 \end{bmatrix}.$$

6.1.6 Calculating Uncertainty Values for Input and Output Data under Polyhedral Uncertainty

We incorporate a perturbation (uncertainty) into the real input data in each dimension and into the output data, after we obtain ***accuracy*** and ***complexity*** terms, to employ our robust optimization technique on the CMARS model. For this purpose, the right-hand side on an uncertainty bound from Eq. (3.16) is evaluated for all input and output values which are represented by *CIs*, and the uncertainty matrices and vectors based on *polyhedral uncertainty* sets are obtained by using Eqs. (3.20) and (3.21).

Furthermore, to perform the given calculations, we need normally distributed data and, since in our dataset some variables are not normally distributed, we use the bootstrapping method of statistics [34], which is the general approach to statistical inference based on building a sampling distribution for a statistic by resampling from the data at hand. With our worst case approach, for the each observation, we use the Eq. (3.16) to receive the uncertainty vectors with their entries u_{km} $(k = 1, 2, \ldots, 38; m = 1, 2, \ldots, 12)$:

$$|u_{k,m}| = |\vartheta_m(\breve{\boldsymbol{x}}_k) - \vartheta_m(\tilde{\boldsymbol{x}}_k)| = \sum_{\substack{A \subseteq \{1,\ldots,K\} \\ \neq}} \mathrm{B}_k^{|A|-1} \prod_{a \in A} \rho_{ka} \cdot \prod_{b \in \{1,\ldots,K\}/A} (\gamma_{kb} + \rho_{kb}). \tag{6.4}$$

Now, we can write our uncertainty matrix for the input data as follows:

$$\boldsymbol{U} = \begin{bmatrix} u_{1,1} & u_{1,2} & \ldots & u_{1,12} \\ u_{2,1} & u_{2,2} & \ldots & u_{2,12} \\ \vdots & \vdots & \ddots & \vdots \\ u_{38,1} & u_{38,2} & \ldots & u_{38,12} \end{bmatrix} \in \begin{bmatrix} [3.5, -3.5] & 0 & \ldots & 0 \\ [3.8, -3.8] & 0 & \ldots & 0 \\ \vdots & \vdots & \ddots & \vdots \\ 0 & [3.2, -3.2] & \ldots & [46.4, -46.4] \end{bmatrix}.$$

After we have incorporated uncertainty for each input value, matrices of our BFs can be expressed in the following forms, just by concentrating on the lower and upper interval boundaries, respectively:

$$\boldsymbol{W}_{up} = \boldsymbol{\vartheta}(\breve{\boldsymbol{b}}) + \boldsymbol{U}_{up} = \begin{bmatrix} 1 & 3.82 & \dots & 0 \\ 1 & 3.82 & \dots & 0 \\ \vdots & \vdots & \ddots & \vdots \\ 1 & 0 & \dots & 47.36 \end{bmatrix},$$

$$\boldsymbol{W}_{low} = \boldsymbol{\vartheta}(\breve{\boldsymbol{b}}) + \boldsymbol{U}_{low} = \begin{bmatrix} 1 & -3.23 & \dots & 0 \\ 1 & -3.79 & \dots & 0 \\ \vdots & \vdots & \ddots & \vdots \\ 1 & 0 & \dots & -45.47 \end{bmatrix}.$$

The output data, the uncertainty vector and the vectors with uncertainty are represented below, respectively:

$$\boldsymbol{v} = \begin{bmatrix} v_1 \\ v_2 \\ \vdots \\ v_{38} \end{bmatrix} \in \begin{bmatrix} [3,-3] \\ [3,-3] \\ \vdots \\ [3,-3] \end{bmatrix}, z_{up} = \breve{\boldsymbol{y}} + \boldsymbol{v}_{up} = \begin{bmatrix} -4.49 \\ -3.56 \\ \vdots \\ -1.87 \end{bmatrix}, z_{low} = \breve{\boldsymbol{y}} + \boldsymbol{v}_{low} = \begin{bmatrix} 1.51 \\ 2.44 \\ \vdots \\ 4.13 \end{bmatrix}.$$

The calculation done above is applicable for both parts of our training dataset.

6.1.7 *Receiving Weak RCMARS Models Using Combinatorial Approach*

As we mentioned in the previous section, PRSS is approximated by a TR problem, and we can easily formulate it as a CQP problem. Moreover, we incorporate a perturbation (uncertainty) into the real input data, $\boldsymbol{x}_k$ $(k = 1, 2, \dots, 38)$, in each dimension and into the output data, $\boldsymbol{y}$, by using our robust optimization approach for a robustification of CMARS. For this aim, by applying Eqs. (3.13) and (3.17) we obtain the uncertainty matrices and vectors based on polyhedral uncertainty. Then, using relation in Eq. (6.4) we evaluate uncertainty for all input and output values which are represented by CIs.

For our example, the uncertainty matrix of input data presented as a vector has a huge dimension ($2^{456(=38\cdot 12)}$) with polyhedral uncertainty, and we do not have enough computer capacity to solve our problem for this matrix. In fact, we have a *tradeoff* between tractability and robustification (cf. Sect. 3.1.3). To overcome that obstacle, in this example, we robustify our CQP problem for each sample value (observation) using the combinatorial approach, which we call *weak robustification*. That weak robustification encounters a data-wise robustification that refers to all the other data according to the interval midpoints ("ceteris paribus"), and it finally addresses the worst case with respect to all the data-wise robustifications. Consequently, we obtain 38 different *weak RCMARS* (*WRCMARS*) models, for each part of our dataset, and solve them with MOSEK [87]. Based on polyhedral uncertainty sets, to solve our problem, we use their vertices. In order to find them, we need especially to apply the Cartesian product of all the intervals of input data in the observations. Hence, our WRCMARS models have different structures depending on the number of entries (BFs), which are used to explain the observations. For instance, we can represent the last observation's WRCMARS model, which has 3 entries, in the following form:

$$\begin{aligned}
&\underset{w,\alpha}{\text{minimize}}\ w,\\
&\text{subject to} \quad 1.51069 - \alpha_0 - 0.29234\alpha_1 - 0.35539\alpha_4 = \beta_1,\\
&\qquad 2.43887 - \alpha_0 - 0.01516\alpha_1 - 0.10152\alpha_3 = \beta_2,\\
&\qquad \vdots\\
&\qquad -1.87353 - \alpha_0 + 2.677\alpha_2 + 3.090\alpha_3 + 45.474\alpha_5 = \beta_{608},\\
&\qquad (\beta_1^2 + \beta_2^2 + \ldots + \beta_{38}^2)^{1/2} \le w,\\
&\qquad (\beta_{39}^2 + \beta_{22}^2 + \ldots + \beta_{76}^2)^{1/2} \le w,\\
&\qquad \vdots\\
&\qquad (\beta_{571}^2 + \beta_{572}^2 + \ldots + \beta_{608}^2)^{1/2} \le w,\\
&\qquad (\beta_{609}^2 + \beta_{610}^2 + \cdots + \beta_{620}^2)^{1/2} \le K^{1/2},
\end{aligned}$$

refering the some $K \geq 0$. In order to solve this problem, we transform it into the MOSEK format above. For this transformation, we attribute new unknown variables in the linear terms which are lying in these 17 cones. By this, in fact, we simplify the notations in the cones and write them as equality and inequality constraints. Therefore, for our last sample, our problem includes 620 linear constraints and 17 quadratic cones.

We write this formulation for each value of our sample ($N = 38$) and solve them separately by using MOSEK program [87]. MOSEK apply an interior-point optimizer, which is an implementation of a homogeneous and self-dual algorithm. We obtain MOSEK results and find the w values for all auxiliary problems; then, using the worst-case approach, we select the solution which has the *maximum* w value. Then, we continue with our calculations using the parameter values α_j ($j = 1, 2, \ldots, 12$) that we find from the auxiliary problem which has the highest w value.

6.1.8 *Sensitivity to the Changes in the Confidence Interval Limits of RCMARS*

In order to represent sensitivity to the changes in the CI limits of the input data and output data and to find suitable interval limits for us, we obtain different uncertainty matrices, $\boldsymbol{U}$, for the input data and different uncertainty vectors, $\boldsymbol{v}$, for the output data as the form of Eq. (3.18) by using 7 different intervals. These ones are given by the pairs $\pm 3, \pm 3/2, \pm 3/4, \pm 3/6, \pm 3/8, \pm 3/10$ and, as a special case, the mid-point value of our interval (i.e., zero lengths interval). In the *latter case*, it reduces to the CMARS model. This shows that CMARS is a *special case* of RCMARS. Therefore, we calculate our parameters with 7 different uncertainty scenarios using these values under polyhedral uncertainty sets for our training data set.

In Sect. 6.1.9, all of the parameter estimates as well as model accuracies for different uncertainty scenarios are shown. When we apply the K values in our RCMARS code and solve it by MOSEK, we use that K value which has the minimum value of PRSS approximately in Eq. (3.22). In order to compare the results concerning accuracy for RCMARS and CMARS, we employ *Average Absolute Error* (*AAE*) and *Root Mean Squared Error* (*RMSE*). Also, we represent variances (σ^2) of CMARS and RCMARS in Sect. 6.1.9.

6.1.9 *Results and Discussion*

In this study, we construct uncertainty matrices, $\boldsymbol{U}$, for the input data and uncertainty vectors, $\boldsymbol{v}$, for the output data and, we recieve 7 different uncertainty scenarios by using the interval values, $\pm 3, \pm 3/2, \pm 3/4, \pm 3/6, \pm 3/8, \pm 3/10$ and zero.

From Tables 6.1 and 6.2 it seems that the solutions obtained are sensitive to the limits of CIs. When the lengths of the CIs are narrow, we evaluate better performance results. Moreover, as in our previous study [98], when we use the *mid-point* (zero value) of our interval values for both input and output data, which is the certain data case; we receive the same parameter estimates as we obtained for CMARS. This is our particular *special case*. When we assess the $\vartheta_m(\boldsymbol{x})$ values in our RCMARS code and employ MOSEK, RCMARS provides us several solutions, each of them based on 12 BFs.

For the training data, models for RCMARS have a smaller variance, but a lower accuracy than CMARS, which is consistent with our expectation. However, we have unexpected results for the testing data.

For the test data and for some suitable uncertainty values, RCMARS produced more accurate model with a smaller variance than CMARS, which can be seen in Table 6.2. This is mainly due to the randomness involved in the input-output variables. According to the above results, we can say that RCMARS can be a more *accurate model* with a **smaller variance** than CMARS.

Table 6.1 Parameter estimates and the model performances for the training data

U, v	±3	±3/2	±3/4	±3/6	±3/8	±3/10	*Zero*
RCMARS							CMARS
α_0	−0.053	0.013	0.135	0.139	0.151	0.139	0.110
α_1	0.078	0.050	−0.040	−0.051	−0.065	−0.063	−0.061
α_2	0.008	0.016	0.009	0.010	0.006	−0.006	−0.024
α_3	−0.045	−0.059	−0.091	−0.103	−0.119	−0.138	−0.139
α_4	−0.021	−0.101	−0.175	−0.166	−0.164	−0.163	−0.155
α_5	0.000	−0.058	−0.113	−0.117	−0.122	−0.124	−0.118
α_6	0.031	0.052	0.066	0.063	0.063	0.072	0.085
α_7	0.054	0.016	−0.018	−0.011	−0.013	−0.007	0.008
α_8	0.216	0.451	0.497	0.470	0.473	0.474	0.453
α_9	−0.003	−0.008	−0.013	−0.007	−0.021	−0.001	0.082
α_{10}	0.001	0.001	0.002	0.002	0.002	0.004	−0.024
α_{11}	−0.002	−0.018	−0.031	−0.022	−0.013	−0.007	−0.066
α_{12}	−0.005	−0.005	−0.004	−0.004	0.006	0.012	0.038
σ^2	0.028	0.057	0.085	0.085	0.092	0.101	0.165
AAE	0.735	0.707	0.678	0.673	0.662	0.656	0.627
RMSE	1.175	1.121	1.078	1.070	1.052	1.037	0.999

Table 6.2 Parameter estimates and the model performances for the testing data

U, v	±3	±3/2	±3/4	±3/6	±3/8	±3/10	*Zero*
RCMARS							CMARS
σ^2	0.005	0.006	0.005	0.005	0.005	0.006	0.012
AAE	0.830	0.831	0.818	0.818	0.812	0.814	0.825
RMSE	1.156	1.163	1.146	1.145	1.138	1.145	0.168

6.2 A Real-World Application of RCMARS in the Energy Sector

Electricity price forecasting models have recently been constructed in Turkey since the electricity market evolved into a competitive form. New market structure is based on a day-ahead price forecasting. Electricity price modeling enables decision makers to see projections for the future. Since the fluctuations in electricity demand affect electricity prices, the prices can change in short-term periods even in a day. Fluctuations in the electricity consumption show that there are **three periods**; *day*, *peak*, and *night*, according to the demand. Therefore, the aim of the study [147] is to make short-term projections for competitive Turkish electricity market where only day-ahead prices are forecasted, and to propose a customized approach for electricity price modeling of Turkey.

Several models are studied in the literature for competitive electricity markets. The categorization of models is based on three main approaches: game theory models, time series models, and production cost models [50]. Commonly, next-day's electricity prices are predicted by using time series models, specifically dynamic regression model [94]. The approach proposed here is based on robust and continuous optimization techniques via our new robust tool, RCMARS. One traditional and one new approaches are proposed and then analyzed considering three different types of period in a day. The results show that with small variance RCMARS performs better than the dynamic regression (DR). Although dynamic regression is not appropriate for small-sized data sets, it is used in order to compare the traditional approach and the customized approaches.

6.2.1 Dynamic Regression Approach

One of the effective methods for price modeling is using a dynamic procedure, since the behavior of the variables over time changes the structure of the price models. The model in Eq. (6.5) is a dynamic regression model that consists of electricity price p_{t+1} at time $t+1$ explained by past prices at times $t, t-1, \ldots, t-k$ and the values of demand at the time $t, t-1, \ldots, t-k$:

$$P_{t+1} = \beta_0 d_t + \beta_1 d_{t-1} + \ldots + \beta_n d_{t-k} + \delta_0 p_t + \delta_1 p_{t-1} + \ldots + \delta_k p_{t-k} + \varepsilon_t, \quad (6.5)$$

where β_i, δ_i represent the coefficients and ε_t stands for the noise terms. This method is used in order to overcome the serial correlation in error [51, 94]. Here, the DR approach is used for the prediction of electricity price in Turkey as a traditional approach. Since the efficiency of the method depends on the selection of explanatory variables, the appropriate model for Turkish electricity market is defined by using the real data set of March 2011. The resulting model is

$$P_{t+1} = \beta_0 d_t + \beta_1 d_{t-7} + \delta_0 p_t + \varepsilon_t. \quad (6.6)$$

Here, the model relates next day's price to current day's demand and price as well as the demand of the same day of the previous week.

6.2.2 CMARS

In order to implement the second step of the algorithm, the MARS models are obtained for each subset by using the Salford MARS System, then the maximum number of BFs M_{max} and the highest degree of interactions are determined. The largest model for the first period, i.e., day, is found to be In order to implement the second step of the algorithm, the MARS models are obtained for each subset by using the Salford MARS System, then the maximum number of BFs M_{max} and the highest degree of interactions are determined. The largest model for the first period, i.e. a day, is found to be

$$\begin{aligned}\hat{y}_d =&\alpha_0 + \alpha_1 \max\{0, x_3 - 0.63\} + \alpha_2 \max\{0, 0.63 - x_3\}+\\ &\alpha_3 \max\{0, x_2 + 2.04\} \max\{0, 0.63 - x_3\}+\\ &\alpha_4 \max\{0, x_1 + 2.7\} \max\{0, x_3 - 0.63\} + \alpha_5 \max\{0, x_1 + 2.7\}+\\ &\alpha_6 \max\{0, x_1 - 0.51\} + \alpha_7 \max\{0, 0.51 - x_1\} + \alpha_8 \max\{0, x_1 + 0.28\}+\\ &\alpha_9 \max\{0, -0.28 - x_1\}.\end{aligned}$$

CMARS algorithm is performed for various values of the bound K to find the minimum PRSS in Eq. (3.6). The model is solved in MATLAB environment for three explanatory variables and the results are given in the Sect. 6.2.4, below.

6.2.3 RCMARS

Electricity price models include uncertain parameters. For instance, small perturbations in electricity price and demand may cause different day-ahead electricity price models. In order to avoid unstable solutions, *all* input and output variables are assumed as random variables, opposite to DR and CMARS, where only the output variable (dependent variable) is regarded as random through noise; now, our RO approach is applied to refering to BFs obtained from MARS. By using Eq. (3.18), uncertainty matrices and vectors for the input and output parameters are constructed based on polyhedral uncertainty sets that are represented by standard confidence intervals. RCMARS model takes its general form with the vector of explanatory variables under uncertainty [12, 96]. To solve the problem, PRSS in Eq. (3.17) is reformulated as a RCQP in Eq. (3.23).

6.2.4 Results and Comparison

Proposed models, CMARS-RCMARS, and the traditional model, DR, are applied to predict day-ahead electricity prices of Turkey. One month is chosen and daily periodic data are used to forecast the electricity prices. Numerical results are represented in Table 6.3 for one period. Here, we consider to present results for only one period (e.g., peak) since the models give similar results for the other two periods (e.g., day and night).

Five different performance measures, namely, *EV* which is our main performance measure, *MAE*, *RMSE*, R^2_{adj} and r, are used to assess the prediction performance of the methods. These measures, their abbreviations, explanations, interpretations and formulas are represented in Table D.1. Moreover, in RCMARS, parameters are evaluated for four uncertainty scenarios using the values under polyhedral

Table 6.3 Comparison of electricity price models based on *AAE*, *RMSE*, R^2_{adj}, *EV* and *r*

	DR	RCMARS1	RCMARS2	RCMARS3	RCMARS4
EV	0.33	0.34	0.007	0.25	0.32
AAE	0.75	0.53	0.82	0.57	0.54
RMSE	0.99	0.86	1.19	0.88	0.86
R^2_{adj}	0.13	0.26	0.42	0.23	0.26
r	0.33	0.73	0.35	0.74	0.73

uncertainty sets. The results are represented in the Table 6.3 with RCMARS1 (CMARS), RCMARS2, RCMARS3 and RCMARS4.

According the results, when RCMARS is applied in Turkish electricity market, better predictions can be received with smaller variance. Also, it can be deduced that RCMARS performs better when the length of confidence intervals is reduced for our performance measures, except EV; it is better when the length of confidence intervals is increased.

6.3 A Real-World Application of RCMARS in the Environmental Sector

6.3.1 Introduction

Climate change has been happening for decades, but it has only recently begun to spark more serious concern due to the severity of the disasters to which it has been attributed. Climate change causes a change in the mean (i.e., the center of location) as well as an increase in the variability (i.e., the spread) of meteorological variables. These changes to the climate might result in, for example, extreme amounts of precipitation occur, which may lead to floods and droughts, which in turn, affect the environment, agriculture and the economy. Thus, the ability to forecast water levels and manage water resources has also gained in significance [104].

Precipitation is a very complicated physical process in nature, which makes it difficult to forecast. Nevertheless, recent positive developments in predictive data mining techniques, which are used in early warning systems [6], are improving the accuracy of precipitation forecasts. This assists in the decision to implement action plans in advance of any predicted potential disaster. The methods used for constructing precipitation models include statistical models, like LR models, splines, time-series models (e.g., ARIMA), computational models, such as Artificial Neural Networks (ANNs) [77], MARS [29], wavelet-ANNs and soft computing models, like neuro-fuzzy and wavelet-neuro-fuzzy models.

Comparison studies reveal that statistical models are not as successful as computational models [74, 77, 95, 107, 134]. The neuro-fuzzy approach performs well, but only when combined with wavelet transforms. Similarly, even though ANNs are used extensively in predicting precipitation, they do not perform well unless they are used in conjunction with another method such as wavelet transformation. MARS considered to be the best performing method compared to the other methods mentioned above [1, 2]. Because of successful track record of the MARS method in precipitation modeling, in the study [104], we attempted our technique based on RCMARS, both in theory and application to be used for the aforementioned purpose. For this goal, a dataset consisting of seven meteorological variables recorded at 43 stations in the continental*Central Anatolia* (*CCA*) region of Turkey over the period 1976–2010 was selected. Details of the dataset studied are presented in the following Sect. 6.3.2.

6.3.2 Dataset and Its Preprocessing

The dataset studied involves seven meteorological variables, namely, the monthly precipitation total (in millimeters), monthly mean temperature, monthly relative humidity (in percent), cloudiness, vapor pressure, surface air temperature, mean pressure and mixing ratio. Here, the mixing ratio is a derived variable obtained as the ratio of (0.622 vapor pressure)/(pressure-vapor pressure) [130]. In addition, time is also considered as *another* independent variable in the model development due to time involvement in the data. The data consists of the values of the above named variables recorded at the 43 stations of the CCA region of the Turkish State Meteorological Service (TSMS) over the period 1976–2010. Note here that the stations taken into account in the study were determined as a result of another study [60, 148].

In RCMARS methodology, since there is a tradeoff between tractability and robustification, we had difficulties regarding computer capacity to solve the optimization problem using uncertainty matrices on a large amount of data, containing seven variables with 420 rows (one for every month in 35 years), for each one of the 43 recording stations. To handle this problem, the size of data was reduced by taking yearly averages of each meteorological variable over all stations. Hence, the dataset was decreased to a size that was more suited to the available computer capacity. Furthermore, the variables were normalized to construct CIs in the interval $[-3, 3]$.

In this application, to compare the performances of prediction models obtained, we also employed the *hold-out method* as the validation technique, where the dataset is divided into two subsamples as *training* and *test* sets. As the dataset incorporates a time series of meteorological variables, it was not subdivided randomly. Instead of this, the first 30 years (from 1976–2005) of each variable regarded were assigned to be the training dataset whereas the last 5 years of the series were assigned to be the test dataset.

6.3.3 Criteria and Measures Used in Performance Evaluations

Our basic performance measure to evaluate the precision of the models was the *variance* and, in this study, it was measured in particular by the *estimation variance* (*EV*). Additionally, to compare the results concerning the accuracies of RCMARS, CMARS and MARS methods, the models developed were further evaluated based on some accuracy measures like R_2, *AAE*, *RMSE* and r. These measures, their abbreviations, explanations, interpretations and formulas are presented in Table D.1. Besides, the models were evaluated with respect to the stabilities of all the measures considered. Here, the stability criterion of a measure compares the performance of a method on both the training and test data. The stable methods are the ones that perform equally well on both training and test datasets.

6.3.4 Developing Precipitation Models

First, using the training dataset described above, several MARS models were developed using Salford System's MARS software [83]. After picking the best one among them, the CMARS model was constructed and robustified under polyhedral uncertainty as described in Sect. 3.1. While developing RCMARS models, a *sensitivity study* was conducted to define the most suitable confidence limits on both the input and output data, $\boldsymbol{x}_k, y_k (k = 1, 2, \ldots, 30)$. For this aim, different uncertainty matrices, $\boldsymbol{U}$, for the input data, $\boldsymbol{x}_k$, and different uncertainty vectors, $\boldsymbol{v}$, for the output data, y_k, were constructed by using four different intervals. These are represented by the pairs $\pm 3/5, \pm 3/10, \pm 3/20$ and 0 (i.e., zero-length interval). Here, the zero-length interval refers to a special case where the RCMARS model reduces to the CMARS model. We estimated our parameters with four different uncertainty scenarios using PRSS values of the Eq. (3.19) under polyhedral uncertainty sets for our training data set (see Table C.1). Here, the values of bound K were determined by a model-free method, and the one having the minimum value of approximate PRSS given in Eq. (3.17) was used.

Owing to the tradeoff between tractability and robustification in RCMARS methodology, difficulties arise that stem from having insufficient computer capacity to solve the RCMARS model using uncertainty matrices and a huge amount of input data (cf. Sect. 3.1.3). To overcome this problem, our combinatorial approach, called *weak robustification* was employed on each sample value (observation) to convert the RCMARS into a CQP problem under polyhedral uncertainty. In the study [104], for each observation, we include perturbation (uncertainty) into the input data, $\boldsymbol{x}_k$, for each dimension, and also in the output data, y_k $(k = 1, 2, \ldots, 30)$, with the help of uncertainty the matrices and vectors constructed according to Eq. (3.18). Thus, 30 different submodels, or *weak* RCMARS (WRCMARS) models, were built as a result. In the WRCMARS algorithm, the MARS models were obtained by using Salford System's MARS software, and then, the maximum number of BFs (M_{max})

and the highest degree of interactions were defined. For this data set, M_{max}, and the highest degree of interaction are assigned to be 12 and 1, respectively. We note that *a main effect model* is developed as a result. Thus, the largest model obtained by the forward MARS algorithm involves the following BF[2]

$$\begin{aligned}
\vartheta_1(\boldsymbol{x}) &= \max\{0, x_2 + 2.0927\}, & \vartheta_2(\boldsymbol{x}) &= \max\{0, x_3 + 1.4227\},\\
\vartheta_3(\boldsymbol{x}) &= \max\{0, x_7 + 0.6001\}, & \vartheta_4(\boldsymbol{x}) &= \max\{0, -0.6001 - x_7\},\\
\vartheta_5(\boldsymbol{x}) &= \max\{0, x_6 - 0.2563\}, & \vartheta_6(\boldsymbol{x}) &= \max\{0, 0.2563 - x_6\},\\
\vartheta_7(\boldsymbol{x}) &= \max\{0, x_5 + 0.0875\}, & \vartheta_8(\boldsymbol{x}) &= \max\{0, -0.0875 - x_5\},\\
\vartheta_9(\boldsymbol{x}) &= \max\{0, x_4 + 2.3288\}, & \vartheta_{10}(\boldsymbol{x}) &= \max\{0, x_1 + 2.4477\},\\
\vartheta_{11}(\boldsymbol{x}) &= \max\{0, X_4 + 0.1409\}, & \vartheta_{12}(\boldsymbol{x}) &= \max\{0, -0.1409 - x_4\}.
\end{aligned}$$

Here, x_1, x_2 and x_3 are the normalized mean temperature, cloudiness, and vapor pressure; x_4 and x_5 are the first-order lagged cloudiness and mean pressure; x_6 and x_7 are the fifth-order lagged cloudiness and vapor pressure, respectively. Hence, the RCMARS model obtained is a 'distributed lag' model due to the fact that it includes lagged independent variables. To prevent nondifferentiability in the optimization problem, the knot values selected are different from but very much close to the corresponding input data. As a result, the largest model can be described as follows:

$$\begin{aligned}
\hat{y} = \alpha_0 + \sum_{m=1}^{M} \alpha_m \vartheta_m(\boldsymbol{x}) &= \alpha_0 + \alpha_1 \vartheta_1(\boldsymbol{x}) + \alpha_2 \vartheta_2(\boldsymbol{x}) + \\
&\quad \alpha_3 \vartheta_3(\boldsymbol{x}) + \alpha_4 \vartheta_4(\boldsymbol{x}) + \alpha_5 \vartheta_5(\boldsymbol{x}) + \alpha_6 \vartheta_6(\boldsymbol{x}) + \alpha_7 \vartheta_7(\boldsymbol{x}) + \\
&\quad \alpha_8 \vartheta_8(\boldsymbol{x}) + \alpha_9 \vartheta_9(\boldsymbol{x}) + \alpha_{10} \vartheta_{10}(\boldsymbol{x}) + \alpha_{11} \vartheta_{11}(\boldsymbol{x}) + \alpha_{12} \vartheta_{12}(\boldsymbol{x}) \\
&= \alpha_0 + \alpha_1 \max\{0, x_2 + 2.09278\} + \\
&\quad \alpha_2 \max\{0, x_3 + 1.4228\} + \alpha_3 \max\{0, x_7 + 0.6002\} + \\
&\quad \alpha_4 \max\{0, -0.6002 - x_7\} + \alpha_5 \max\{0, x_6 - 0.2564\} + \\
&\quad \alpha_6 \max\{0, 0.2564 - x_6\} + \alpha_7 \max\{0, x_5 + 0.0876\} + \\
&\quad \alpha_8 \max\{0, -0.0876 - x_5\} + \alpha_9 \max\{0, x_4 + 2.3289\} + \\
&\quad \alpha_{10} \max\{0, x_1 + 2.4478\} + \alpha_{11} \max\{0, x_4 + 2.76403\} + \\
&\quad \alpha_{11} \max\{0, -0.1410 - x_4\}.
\end{aligned}$$

[2]For the ease of representation, here and subsequently, we suppress the index m of the subvectors $\boldsymbol{x}^m$ and just write $\boldsymbol{x}$.

Thirty different submodels were solved individually by using the MOSEK program and, thus, the w values were determined for all auxiliary problems. Then, using the worst-case approach, the solution chosen was the one with maximum w value, and the parameters α_j ($j = 1, 2, \ldots, 12$) were estimated (see Table C.1).

6.3.5 Results and Discussion

The models developed as defined in the previous section, were evaluated with respect to the criteria by using the formulas represented in Table D.1. The results are given in Table C.2. According to them, the following findings can be indicated.

- For $U = \pm 3/5$, the best measure values for training data were constructed for $v = \pm 3/20$ other than PV measure; it was best for $v = \pm 0$.
- For $U = \pm 3/10$ and $\pm 3/20$, the same best values for training data were obtained for $v = \pm 3/20$.
- For $U = \pm 3/5$, $\pm 3/10$ and $\pm 3/20$, the best values for test data were obtained for $v = \pm 0$.
- For $U = \pm 3/5$, the best values for the stabilities of measures were evaluated for $v = \pm 3/20$, whereas for $U = \pm 3/10$ and $\pm 3/20$, the best values for the stabilities of measures were calculated for $v = \pm 0$.
- For $U = \pm 0$, all measures were the same for the training, test and stabilities.
- The best values for the training data were received for $U = \pm 3/10$ or $\pm 3/20$ and $v = \pm 3/20$.
- The best values for test data were constructed for $U = \pm 3/5$ and $v = \pm 0$, while the best stabilities of measures were obtained for $U = \pm 3/10$ or $\pm 3/20$ with $v = \pm 0$.

Based on the above findings, the best RCMARS solution was determined for $U = \pm 3/5$, $\pm 3/10$ or $\pm 3/20$ and $v = \pm 0$. For the goal of comparison, we took $U = \pm 3/10$ or $\pm 3/20$ and $v = \pm 0$. The performance measures of MARS, CMARS and RCMARS are given in Table 6.4. Note: * indicates the best performance for train, test and stability (st), with respect to the corresponding performance measure.

The results implied the following conclusion:

Table 6.4 Performance measures of the precipitation models

	MARS			CMARS			RCMARS		
	Train	Test	St	Train	Test	St	Train	Test	St
R^2	0.957	0.139	0.145	0.971*	0.225	0.231	0.876	0.789*	0.901*
AAE	0.165	0.701	0.235	0.131*	0.6463	0.203	0.273	0.311*	0.877*
RMSE	0.204	0.830	0.246	0.166*	0.788	0.211	0.346	0.411*	0.842*
r	0.978	0.652	0.666	0.986*	0.672	0.680	0.950	0.900*	0.947*
EV	0.957	1.324	0.723	0.953	1.241	0.768	0.628*	0.687*	0.914*

- For the training data, CMARS performed better than the other two methods with regard to all measures except EV; it was the best for RCMARS.
- For the test data and stabilities, RCMARS considerably outperformed the other two methods with respect to all measures.

6.4 A Real-World Application with RCGPLM in the Financial Sector

6.4.1 Introduction

In recent years, sovereign debt-servicing difficulties and outright defaults have been observed more frequently than before even though the macroeconomic misalignments causing debt crises are still not well understood. In order to forecast several kinds of crises, the literature has focused on especially 'twin' currency and banking crises, but, not on the prediction of sovereign debt crises. Sovereign debt crises usually occur as the result of outright default on domestic and external debt to rollover/liquidity crises, when investors of a country, which is solvent, but illiquid and also on the verge of default on its debt, are unwilling to roll over short-term debts coming to maturity. Since several countries have large debt burdens and can be subject to debt-servicing problems in the foreseeable future, assessing and forecasting debt sustainability has great empirical and policy importance [33, 82]. In addition to these, internationalism and integration of economies are also essential factors of country risk.

Especially, decision makers and investors should expect the coming risks in the international area to make decisions, take measures and make profitable investments in the right places all over the world. Through the world, emerging markets draw attention due to their high growth potential and high profit expectancies. On the other hand, they are relatively higher risky markets because of volatility of economic policies, weak banking sector, high dependence on external capital flows and uncertain growth prospects. Therefore, they are more prone to the crises [76]. As a classification tool, Logistic Regression models and algorithms are often applied to predict defaults/nondefaults or success/unsuccess, developed by using maximum likelihood method. Although they do not have assumptions like normality and linearity, they have some deficiencies, especially, in correlated variables and incomplete datasets [33, 76].

In the previous study [137], unlike Logistic Regression, the datasets which include both linear and nonlinear variables, are tried to be explained efficiently using a semiparametric model: *CGPLM* (cf. Sect. 3.2.2). Here, it is constructed as a combination of a discrete model of Logistic Regression and a continuous model of CMARS. Comparing CMARS and CGPLM, it is clearly seen that CGPLM has an advantage in terms of reducing the complexity and increasing the rate of accuracy in the results.

In the work [102], we represent a newly developed RCGPLM with a real-world application in finance to predict the default probabilities in 45 emerging markets. In RCGPLM, the linear part consists of a discrete regression model Logistic Regression and the nonlinear part consists of a continuous regression model, RCMARS. The aim of RCGPLM is to decrease the complexity of RCMARS, reducing the number of variables by transferring the linear ones to Logistic Regression. This section employs RCGPLM with a variety of macroeconomic factors to assess affection on the risk of sovereign default and on a debt crisis, for a large sample of countries.

6.4.2 Data

In the application part of the model, we used the same data set as in our previous study [30, 137], where we employed Conic Generalized Partial Linear Model, to have a chance to compare the results of the two models. The data set used in this study is quoted, originally, from Fioramanti's paper [43], and it is comprised of some important macroeconomic determiners of debt crises in 45 emerging markets between the years 1980 and 2005. The time-series data contain 1019 observations with a dependent variable that shows whether the country is in a debt crisis taking the value '0' (non-default) or the value '1' (default) values, and with 13 independent variables:

X_1 : Bank liquid reserves to bank assets ratio,

X_2 : Changes in net reserves/GDP (Gross Domestic Product),

X_3 : Current account balance (% of GDP),

X_4 : Exports of goods and services (% of GDP),

X_5 : External debt total/Total Reserves,

X_6 : Long-term debt/GDP,

X_7 : GDP growth (annual %),

X_8 : Liquid liabilities as % of GDP,

X_9 : Total debt servic (% of exports of goods services and income),

X_{10} : Short-term debt (% of exports of goods services and income),

X_{11} : Trade (% of GDP),

X_{12} : Use of IMF credit/GDP,

X_{13} : Inflation consumer prices (annual %).

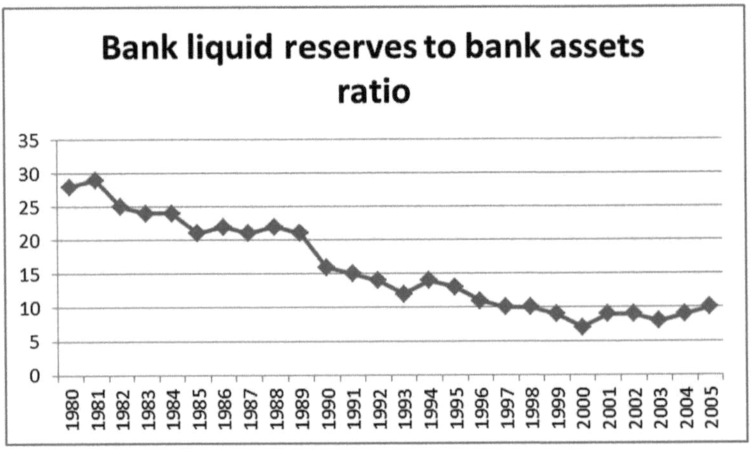

Fig. 6.1 Graph for the character of X_1

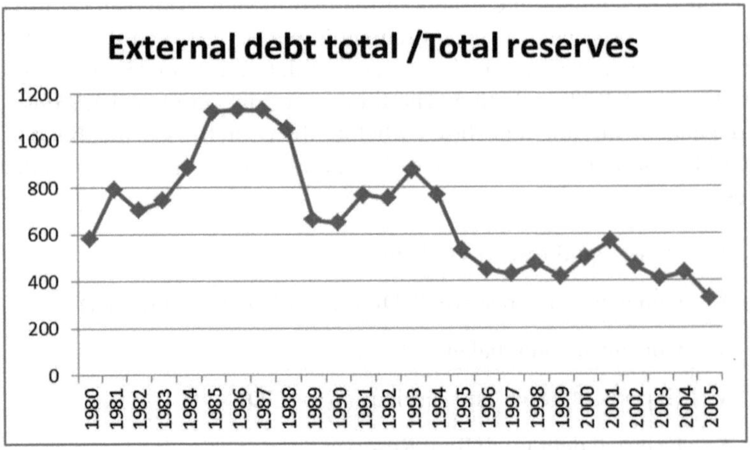

Fig. 6.2 Graph for the character of X_5

In Figs. 6.1, 6.2, 6.3, 6.4, 6.5, we can see the character of some of our variables belonging to a selected country among 45 countries, for a visualization of the dataset described above.

Our *training sample* is beneficial to construct the model based on 757 observations which belong to the years 1980–1999, while a *testing (validation) sample* is used to test the model including 262 observations which belong to the years 2000–2005. Here, to overcome the capacity problem in MATLAB, we need again bootstrapping to reduce the number of observations to an applicable number and to conserve all specific properties of the data.

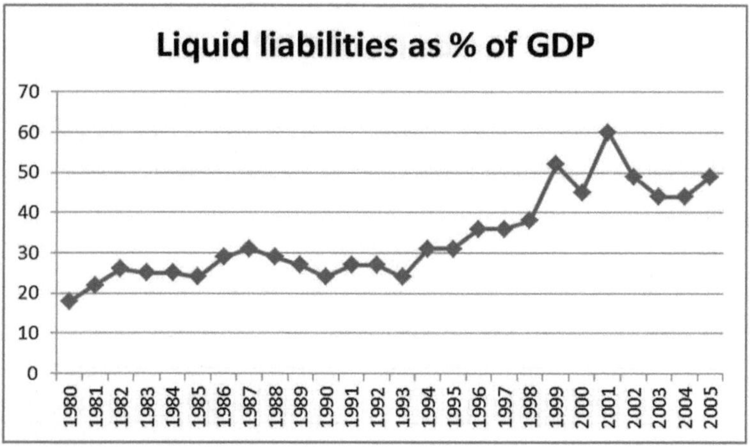

Fig. 6.3 Graph for the character of X_8

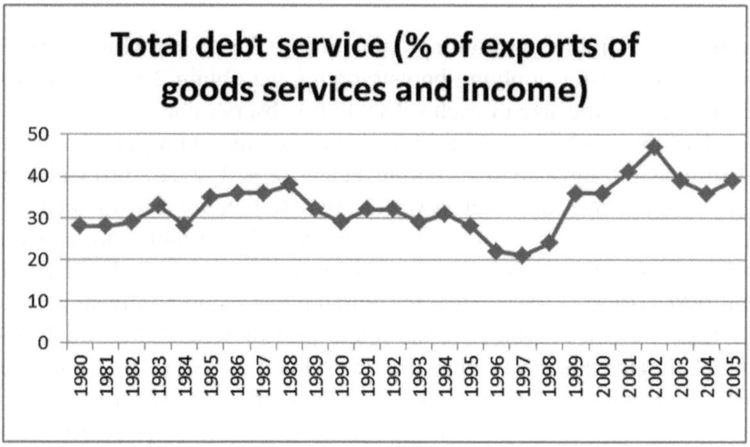

Fig. 6.4 Graph for the character of X_9

6.4.3 Application

Derivation of the model from the *training sample*:

To predict the default probabilities of emerging markets, we use a large sized real-world financial data as an application of RCGPLM. In our methodology, we use a *tradeoff* between tractability and robustification leading us to a difficulty about computer capacity to solve the problem equipped with uncertainty matrices and a huge size input data. Therefore, on each sample value (observation) in the linear and nonlinear parts, our combinatorial approach, *weak robustification* is applied to convert the RCGPLM into a CQP problem. In addition to that, to overcome

Fig. 6.5 Graph for the character of X_{11}

this problem, we divide the training data set into 2 subsets which have 378 and 379 observations. After applying bootstrapping, we obtain 2 normally distributed samples and reduce the size of each subset to 60 observations.

On each of these subsets, we insert perturbation (uncertainty) in the input data x_k for each dimension, and also in the output data y_k ($k = 1, 2, \ldots, 60$) with the help of the uncertainty matrices and vectors which are based on *polyhedral uncertainty sets* constructed in Eqs. (3.20) and (3.21). In that way, the variables are converted into standard normal distribution to obtain CIs in the interval $[-3, 3]$. Then, different so-called *WRCGPLMs* [101] appear for both the linear and nonlinear parts.

After that, we can continue on the linear part of RCGPLM, which is defined in Sect. 3.2.4. The linear variables are determined as: X_8 (liquid liabilities as % of GDP), X_9 (total debt service: % of exports of goods services and income), X_{12} (use of IMF credit/GDP) and X_{13} (inflation consumer prices) which have a linear relationship with the dependent variable 'Y'. Then 757 different models are constructed to constitute our WRCGPLM. After the solution of these models in MOSEK and finding the w_1 values for all auxiliary problems, we obtain the solutions which have the *maximum* w_1 value with respect to the Eq. (3.45), herewith applying the *worst-case* approach. The linear least-squares system $\boldsymbol{X\beta}^{preproc}$, where $\boldsymbol{\beta}^{preproc}$ is the optimal vector of the regression and $\boldsymbol{X}$ is the design matrix, is subtracted from the response $\boldsymbol{y}$ to derive the vector $\boldsymbol{\gamma}$ of the nonlinear model (for closer details see the procedure in Sects. 3.2.2, 3.2.5 and 3.2.6.2). As a result, the final regression model can be expressed as in Eq. (3.30).

To prevent from any damage to the binary structure of the dependent variables $\boldsymbol{\gamma}$ in Eq. (3.31), which employs a subtraction of the results from the original $\boldsymbol{y}$ values, we separate the data set into Group I and Group II. Group I consists of the observations giving a result of '0' after the linear regression, while Group II comprised of the observations giving a linear regression result of '1'. From now on, the nonlinear process will be separately applied on these 2 groups of each bootstrapped subsets with the binary residual vector $\boldsymbol{\gamma}$. Then, we construct the largest model for Group I and Group II by using the Salford MARS. For example, the largest model includes the following BFs for Group I[3]:

$$\vartheta_1(\boldsymbol{t}) = \max\{0, t_2 + 1.597\}, \qquad \vartheta_2(\boldsymbol{t}) = \max\{0, t_7 + 1.798\},$$
$$\vartheta_3(\boldsymbol{t}) = \max\{0, t_3 + 1.395\}, \qquad \vartheta_4(\boldsymbol{t}) = \max\{0, t_6 + 1.529\},$$
$$\vartheta_5(\boldsymbol{t}) = \max\{0, t_1 + 2.764\}.$$

Thus, the large model is represented as follows:

$$\begin{aligned}\hat{y} &= \alpha_0 + \sum_{m=1}^{5} \alpha_m \vartheta_m(\boldsymbol{t}) + \epsilon = \alpha_0 + \alpha_1 \vartheta_1(\boldsymbol{t}) + \alpha_2 \vartheta_2(\boldsymbol{t}) + \alpha_3 \vartheta_3(\boldsymbol{t}) + \alpha_4 \vartheta_4(\boldsymbol{t}) + \alpha_5 \vartheta_5(\boldsymbol{t}) \\ &= \alpha_0 + \alpha_1 \max\{0, t_2 + 1.597\} + \alpha_2 \max\{0, t_7 + 1.7978\} + \alpha_3 \max\{0, t_3 + 1.395\} + \\ &\quad \alpha_4 \max\{0, t_6 + 1.529\} + \alpha_5 \max\{0, t_1 + 2.764\}.\end{aligned}$$

On the nonlinear part of the model, our RO technique is employed inserting perturbation (uncertainty) in the real input data $\boldsymbol{t}_k$, in each dimension, and into the output data γ_k $(k = 1, 2, \ldots, 60)$. To reach this goal, similarly to the linear part, CIs are defined for all input and output values by the help of the uncertainty matrices and vectors, which are based on *polyhedral uncertainty sets*, obtained by Eqs. (3.20) and (3.21).

Subsequently, as we did in the linear part, we derive 60 different WRCGPLMs for the nonlinear part. Among the solutions, which are the w_2 values for all auxiliary problems in Eq. (3.48), found in MOSEK program, we decide the optimum solution which has the *maximum* w_2 value in the Eq. (3.48) with the *worst-case* approach (see Sect. 3.2.5 for more details). From now on, the calculations will be completed with the parameter vector $\boldsymbol{\alpha}$ which is obtained from the auxiliary problem with the highest w_2 value.

6.4.4 Application of the Model on the Testing Sample

In this part, the methodology how to measure the effectiveness of the RCGPLM model on the validation sample is discussed. From the training sample, 4 models have been derived by 4 sets which are constructed according to the linear regression results of 2 bootstrapped samples. Firstly, the testing sample is separated into 2 groups each of which exists of 131 observations. Then, to provide the integrity of

[3]For the ease of representation, here and subsequently, we suppress the index m of the subvectors $\boldsymbol{t}^m$ and just write $\boldsymbol{t}$.

Table 6.5 Results of RCGPLM

	Training sample			Validation sample		
	D		ND	D		ND
Default	87.80 %		3.80 %	96.88 %		10.29 %
Non-Default	12.20 %		96.20 %	3.13 %		89.71 %
CCR		93.33 %			92.00 %	

Table 6.6 Comparison of results of CGPLM and RCGPLM

Training sample			Validation sample			
D-D	ND-ND	CCR	D-D	ND-ND	CCR	
LR	69.97 %	87.20 %	79.39 %	82.35 %	89.10 %	87.79 %
CMARS	81.89 %	92.58 %	87.64 %	86.27 %	87.10 %	86.94 %
CGPLM	90.09 %	93.24 %	91.81 %	86.27 %	90.05 %	89.31 %
RCGPLM	87.80 %	96.20 %	93.33 %	96.88 %	89.71 %	92.00 %

application with the training sample, bootstrapping method is employed to reduce the number of observations to 60.

For the linear part, on each of these 60-membered subsets, the linear regression parameters are employed on the linear variables T_k $(k = 1, 2, \ldots, 4)$ to determine '0' and '1' results and to separate any subset into Group I and Group II. For each counterpart of the training subsets, previously obtained RCMARS models and parameters are implemented on the nonlinear variables X_j $(j = 1, 2, \ldots, 9)$. The final output of the model is achieved by summing up the results of our linear and nonlinear parts. However, RCMARS results are standardized to be able to provide the correspondence with the linear regression results which are situated around '0'. For further details, we refer to [102, 137]. The results of this application can be seen from Tables 6.5 and 6.6. Note: CRR indicates Correct classification rate.

6.4.5 *Results and Comparison*

In Table 6.5, we present the obtained numerical results of RCGPLM for training and validation sample. Table 6.6 explains the comparison of the results of Logit Regression, CMARS, CGPLM and RCGPLM. Here, D-D and ND-ND show the crisis and non-crisis situations, which our model predicts truly, respectively. As it can be seen in Table 6.6, RCGPLM provides a 93.33 % accuracy rate, while Logit Regression, CMARS and CGPLM give 79.39 %, 87.64 %, 91.81 %, respectively, for our training data set. Similarly, for the validation data set, we have 92 % accuracy rate for RCGPLM, whereas Logit Regression, CMARS and CGPLM result with 87.79 %, 86.941 %, 89.31 % accuracy, respectively.

In fact, RCGPLM provides better results for both training and validation samples in terms of accuracy rates. In the training sample, RCGPLM expects 87.80 % of crises and 96.20 % of non-crisis situations in emerging markets, giving a total 93.33 % accuracy rate. For our validation sample, the model forecasts 96.88 % of debt crises and 89.71 % of non-crisis situations emerging markets, giving a total 92 % accuracy rate. Here, our *variance values* are 0.0513 for training data and 0.0935 for testing data. With a smaller variance, models for RCGPLM have a higher accuracy than models of Logit Regression and CMARS, and it is considerable higher than models for CGPLM over both the training and validation data. Similarly, regarding the validation sample, the accuracy rate increases. This means that RCGPLM is a functional methodology in datasets of noisy variables with a possibly higher accuracy rate and, in particular, a **smaller variance**.

Chapter 7
Conclusion and Outlook

The great national and international crisis which resulted after the earthquake and tsunami in Japan in 2011 disclosed again the high interdependence of environmental, technological and economical states, and it underlined the necessity for an essential restructuring of the approach to risk and regulation in these areas to cope with uncertain data. Consequently, core elements of a new global regulatory framework have to be established in order to make these systems more robust and suitable for serving the requirements of the real life. Thus, robust optimization has a great importance as a modeling framework for immunizing against parametric uncertainties, and the integration of uncertain data is of considerable importance for the reliability of any model of a highly interconnected system.

In this book, R(C)MARS is worked on in theory and application by important Robust Optimization, and a time-dependent counterparts of R(C)MARS has been further extended and proved to be a general framework of multi-modal regulatory systems under polyhedral uncertainty in this respect. Because of the computational effort which R(C)MARS easily needs, we also describe our new concept of a weak robustification that is called as WR(C)MARS. We study on R(C)MARS in terms of polyhedral uncertainty. This brings us back to CQP naturally. Through R(C)MARS we are also permitted to involve uncertainty in the input variables to regression and classification within modeling; that uncertainty is typical for real-world challenges, too. By conducting a robustification in (C)MARS, we aim to reduce the estimation variance. In RMARS and RCMARS, however, we have an extra problem to solve (by Software MARS, etc.), namely the knot selection (which is not needed for the linear part). Therefore, we analyze GPLMs, and introduce a newly developed CGPLM and R(C)GPLM, involving the contribution of (C)MARS and R(C)MARS. As semiparametric models, CGPLM and RCGPLM lead to reduce the complexity of (C)MARS and (R)CMARS, that is given by the number of variables used in (C)MARS and R(C)MARS algorithm. In RCMARS, we imply the integral terms as a 'complexity', too.

A. Özmen, *Robust Optimization of Spline Models and Complex Regulatory Networks*, Contributions to Management Science,
DOI 10.1007/978-3-319-30800-5_7

We analyze the regression models of regulatory systems when the entries of the regulatory network are splines as an advanced case, using (C)MARS on parameter estimation for TE networks. We also apply our methods of R(C)MARS in the case of the existence of noise in the expression data which translates into the model, and thus employing robust optimization. In fact, here, the states of target and environmental items depend on uncertain states of target and environmental factors. The prediction of the TE regulatory networks and the following comparison with the underlying data leads to an analysis of regression and classification models for parameter estimation. As an advanced approach to obtain a more flexible model, we consider regression problems for TE regulatory systems when the entries of the regulatory network are splines, and we derive a corresponding robust counterpart program under polyhedral uncertainty. We have introduced a new implementation area of R(C)MARS by a dynamical modeling of regulatory networks, which also include eco-finance networks and gene-environment networks. R(C)MARS method is able to deal with uncertainty in data and, thus, it is a more realistic alternative to modeling of real-life data.

In the book, we briefly review on theory and methods of R(C)MARS and R(C)GPLM. We also conduct applications on data in further areas such as the sectors of energy, finance, biotechnology and ecology. We run the corresponding code for different kinds of data that include uncertainties and, then, evaluate the results with respect to accuracy and stability. Next, the results of the accuracy and sensitivity analysis on the parameter estimates and, thus, the model performances are presented. We solve our optimal problem of R(C)MARS and R(C)GPLM by using the continuous RO approach and a combinatorial variety of them, the weakly robust case, to handle uncertainties that may exist in data and to make our rich approach meaningful and sustainable. In this way, we aim to decrease the estimation variance. Results indicate that for the training data, R(C)MARS models have smaller variances but slightly lower accuracies than (C)MARS models; here, this finding is consistent with our expectation. However, for the testing data and for some suitable uncertainty values, R(C)MARS produced more accurate models with smaller variances than (C)MARS. In the particular application of precipitation forecasting, the RCMARS model developed is twice as much accurate as MARS and CMARS models with respect to MAE and RMSE measures, and it is twice as precise as MARS and CMARS models with respect to prediction variance measure. Furthermore, it has a considerably high stability when compared to those of other two models. To conclude, it can be said that both R(C)MARS produce the best model for the data studied when compared to the MARS and CMARS with respect to precision and stability.

According to all the aforementioned computations and comparisons, our R(C)MARS methods prove to be very competitive with the other methods. We are able to achieve a variance reduction, which is very important in practice and an additional advantage, especially, when comparing with our predecessor method of MARS. Given the existence of uncertainty and noise in real-world data, R(C)MARS and R(C)GPLM model approaches gain importance to reduce complexity and variance of estimation. In future studies, we will investigate on real-world applications of these approaches in some areas, such as regulatory network systems, like gene-environment and eco-finance networks, quality management, biotechnology and financial forecasting, to validate and to investigate the performance of our R(C)MARS and R(C)GPLM.

In all these studies, although we have small datasets for our applications, the uncertainty matrices for the input data have huge dimensions, and we have not had enough computer capacity to solve our problems for those uncertainty matrices. Indeed, we have a tradeoff between tractability and robustification. To overcome this difficulty, we obtain different WR(C)MARS models for all sample values (observations) applying a combinatorial approach, and solve them by running our code and using MOSEK program. In our future studies, we will discuss about how we can obtain a more robust model using different methods and about what further research will consist of in this respect. We plan to also apply parallel computing to solve our problem with the computer capacity.

In our investigated version of R(C)MARS, for convenience, the polyhedral type of uncertainty and normally distributed data are assumed. Obviously, these assumptions lead to some weaknesses on R(C)MARS modeling. In our future studies, ellipsoidal uncertainty will be considered since it uses a more realistic assumption, which leads to a more robust approximation, although it may cause an increased model complexity. Distributional assumptions other than normal or robust estimators may also be considered in the construction of confidence intervals.

In Chap. 2, some background information about multi-model regulatory networks, optimization and regression is given. Theory and approaches of R(C)MARS and R(C)GPLM method under polyhedral uncertainty are presented in Chap. 3. Then, in Chap. 4, spline regression models for complex multi-model regulatory networks are introduced in theory and methods. (C)MARS results based on different datasets for the simulation are also demonstrated in this chapter. In Chap. 5, RO for spline regression models of multi-model regulatory networks are introduced in theory and methodology. R(C)MARS results with different uncertainty scenarios for our numerical example are also studied here. Real-world applications from different sectors are represented in Chap. 6. Finally, the conclusion and outlook to further studies are stated in Chap. 7.

Appendix A
Coefficients and Performance of MARS-CMARS Models for TE Networks

See Tables A.1, A.2, and A.3.

Table A.1 For targets and environmental factors: parameter values of MARS algorithm through 5 different simulated datasets

	α_0	α_1	α_2	α_3	α_4	α_5	α_6	α_7	α_8	α_9
$\tilde{x}_1$	−0.982			2.458		1.193		−1.593		1.191
$\tilde{x}_2$	0.396		1.071	1.269		1.516				
$\tilde{e}_1$	−1.244		0.661		0.425	0.471				
$\tilde{e}_2$	1.763		−1.553	−0.729						
$\tilde{x}_1$	−1.020		0.992		0.484				1.687	
$\tilde{x}_2$	0.370		−0.910	1.193		−0.747				1.962
$\tilde{e}_1$	−2.094	0.608	1.074		0.591		−0.880			
$\tilde{e}_2$	−0.454	1.190	−0.527	−0.688						
$\tilde{x}_1$	−1.915		0.588	0.512						
$\tilde{x}_2$	0.340	−2.115	2.409		−0.126		−0.838			
$\tilde{e}_1$	1.377		−2.596	−0.815	−0.753		1.094	1.077		
$\tilde{e}_2$	−0.085		−0.844	1.409	−0.460	−1.485		1.029		
$\tilde{x}_1$	−0.217				0.604	−1.022	0.600			
$\tilde{x}_2$	−0.355	−0.740			0.450			−0.281		
$\tilde{e}_1$	2.916	−1.433	−1.198			−0.802	−0.864	1.415		
$\tilde{e}_2$	−1.744			0.802	0.312		0.460			
$\tilde{x}_1$	1.087	−0.648		−1.010	0.886					
$\tilde{x}_2$	−0.337		−0.545					0.833		
$\tilde{e}_1$	−0.768				0.843	−0.481			−0.433	
$\tilde{e}_2$	−0.661	2.019	−0.592	0.680						

A. Özmen, *Robust Optimization of Spline Models and Complex Regulatory Networks*, Contributions to Management Science,
DOI 10.1007/978-3-319-30800-5

Table A.2 For targets and environmental factors: parameter values of CMARS algorithm through 5 different simulated datasets

	α_0	α_1	α_2	α_3	α_4	α_5	α_6	α_7	α_8	α_9	α_{10}	α_{11}
$\tilde{x}_1$	−1.690	0.301	−0.238	1.362	1.280	1.011	0.266	−0.342	−2.311	0.622	0.296	
$\tilde{x}_2$	−0.854	−0.075	−0.174	1.076	−0.254	0.438	0.737	0.397	1.408	−0.610	−0.243	−0.122
$\tilde{e}_1$	−3.525	0.200	0.986	−0.557	0.416	0.605	2.114	0.463	−0.259	−2.284	0.499	−0.226
$\tilde{e}_2$	1.618	3.071	−1.217	−1.216	−1.236	0.191	−1.093	0.140				
$\tilde{x}_1$	−3.545	1.489	0.938	0.486	0.733	−0.494	−0.090	2.006	−0.046	1.767	−0.519	−0.835
$\tilde{x}_2$	−1.789	0.037	0.012	1.150	−0.542	−0.946	0.295	0.310	0.885	1.049	0.703	
$\tilde{e}_1$	−1.931	0.385	1.169	−0.027	0.502	−0.578	−1.133	0.343	0.248			
$\tilde{e}_2$	−0.839	1.216	−0.282	−0.658	0.680	−0.224	−0.186	−0.398				
$\tilde{x}_1$	−2.253	0.249	1.096	0.307	−0.332	0.074	0.522			−2.253		
$\tilde{x}_2$	0.418	−2.430	2.113	0.082	−0.272	0.130	−0.701	0.478	0.133	0.418		
$\tilde{e}_1$	1.323	0.451	−2.712	−0.843	−1.019	0.512	1.216	0.985		1.323		
$\tilde{e}_2$	−0.069	−0.578	−1.188	1.462	−0.412	−2.311	0.231	1.615		−0.069		
$\tilde{x}_1$	0.238	−0.067	−0.609	−0.553	0.559	−1.222	0.438	0.484	0.324	1.209	−0.123	−0.426
$\tilde{x}_2$	−0.083	−0.618	−0.748	0.192	0.557	−0.361	−0.353	−0.267	−0.083			
$\tilde{e}_1$	1.460	−0.197	−0.891	−0.496	−0.455	−0.806	−0.700	0.377	0.796	−0.289	−0.006	
$\tilde{e}_2$	−1.721	−0.101	0.786	0.781	0.150	0.423	0.246	0.196				
$\tilde{x}_1$	0.644	−0.464	0.262	−0.881	0.807	−0.045	0.388	0.008				
$\tilde{x}_2$	−0.881	−0.106	−0.174	−0.095	0.216	0.311	−0.060	0.501				
$\tilde{e}_1$	0.037	−0.330	−0.093	−0.101	0.645	−0.345	−0.278	−0.170	−0.400			
$\tilde{e}_2$	0.543	0.343	−0.961	−0.327	1.287	0.043	−0.255	0.629	−1.426	0.522	0.664	1.186

Table A.3 Performance measures of MARS and CMARS models based on 5 different simulated datasets

		MARS				CMARS			
		$\tilde{x}_1$	$\tilde{x}_2$	$\tilde{e}_1$	$\tilde{e}_2$	$\tilde{x}_1$	$\tilde{x}_2$	$\tilde{e}_1$	$\tilde{e}_2$
1	R^2_{adj}	0.8639	0.9004	0.8639	0.8167	0.8835	0.9005	0.9064	0.8897
	AAE	0.2474	0.2398	0.2178	0.3437	0.1911	0.1870	0.1750	0.2573
	$RMSE$	0.3124	0.2976	0.3159	0.4383	0.2419	0.2340	0.2061	0.2989
	r	0.9416	0.9554	0.9386	0.9121	0.9654	0.9727	0.9743	0.9601
2	R^2_{adj}	0.9017	0.9481	0.9415	0.8946	0.9057	0.9607	0.9652	0.9523
	AAE	0.2059	0.1773	0.1859	0.2366	0.1330	0.1203	0.1276	0.1587
	$RMSE$	0.2662	0.2134	0.2140	0.3261	0.2052	0.1553	0.1476	0.1974
	r	0.9560	0.9781	0.9753	0.9528	0.9741	0.9885	0.9883	0.9830
3	R^2_{adj}	0.8601	0.9422	0.8532	0.9574	0.9018	0.9426	0.9289	0.9654
	AAE	0.1609	0.0911	0.1460	0.0924	0.1141	0.0803	0.0850	0.0883
	$RMSE$	0.2147	0.1267	0.1921	0.1369	0.1627	0.1129	0.1265	0.1167
	r	0.9337	0.9756	0.9401	0.9830	0.9625	0.9807	0.9745	0.9877
4	R^2_{adj}	0.8905	0.9559	0.8110	0.9345	0.9315	0.9837	0.8141	0.9525
	AAE	0.1694	0.1413	0.1940	0.1687	0.1095	0.0891	0.1664	0.1273
	$RMSE$	0.2088	0.2027	0.2328	0.2100	0.1299	0.1075	0.1982	0.1608
	r	0.9509	0.9805	0.9222	0.9709	0.9813	0.9945	0.9442	0.9830
5	R^2_{adj}	0.9084	0.9162	0.9025	0.9125	0.935	0.9512	0.95	0.9605
	AAE	0.1658	0.1700	0.1554	0.2007	0.1227	0.1133	0.095	0.0935
	$RMSE$	0.2043	0.2314	0.1958	0.2508	0.1547	0.155	0.122	0.1274
	r	0.9591	0.9608	0.9564	0.9609	0.9768	0.9825	0.983	0.9901

Appendix B
Performance of R(C)MARS Models for TE Networks

See Tables B.1 and B.2.

Table B.1 Performance measures of LR, MARS and RCMARS models based on different simulated data for each target

x_1						x_2						
	LR	MARS	RCMARS				LR	MARS	RCMARS			
EV	0.736	0.938	0.562	0.666	0.817	0.852	0.871	0.917	0.697	0.748	0.819	0.866
R^2_{adj}	0.684	0.925	0.761	0.840	0.926	0.940	0.859	0.901	0.824	0.860	0.902	0.925
AAE	0.333	0.175	0.317	0.257	0.169	0.151	0.231	0.189	0.220	0.195	0.173	0.161
RMSE	0.503	0.244	0.379	0.31	0.211	0.190	0.336	0.282	0.335	0.299	0.251	0.219
r	0.858	0.968	0.942	0.959	0.979	0.983	0.939	0.958	0.946	0.957	0.969	0.976
	LR	MARS	RCMARS				LR	MARS	RCMARS			
EV	0.666	0.869	0.431	0.563	0.748	0.864	0.794	0.930	0.584	0.721	0.911	0.926
R^2_{adj}	0.598	0.843	0.65	0.75	0.845	0.896	0.753	0.916	0.754	0.831	0.917	0.920
AAE	0.376	0.268	0.375	0.305	0.244	0.181	0.314	0.187	0.255	0.212	0.166	0.163
RMSE	0.567	0.354	0.488	0.412	0.325	0.266	0.444	0.259	0.371	0.308	0.216	0.212
r	0.816	0.932	0.901	0.924	0.947	0.963	0.891	0.965	0.942	0.955	0.976	0.976
	LR	MARS	RCMARS				LR	MARS	RCMARS			
EV	0.776	0.939	0.519	0.726	0.906	0.92	0.797	0.904	0.582	0.717	0.842	0.856
R^2_{adj}	0.733	0.926	0.734	0.822	0.927	0.938	0.757	0.891	0.758	0.818	0.892	0.902
AAE	0.303	0.193	0.341	0.259	0.153	0.143	0.305	0.211	0.270	0.223	0.181	0.179
RMSE	0.462	0.243	0.413	0.338	0.217	0.200	0.441	0.303	0.406	0.352	0.272	0.259
r	0.882	0.969	0.931	0.943	0.975	0.979	0.893	0.951	0.924	0.938	0.962	0.965

(continued)

A. Özmen, *Robust Optimization of Spline Models and Complex Regulatory Networks*, Contributions to Management Science,
DOI 10.1007/978-3-319-30800-5

Table B.1 (continued)

x_1							x_2					
	LR	MARS	RCMARS				LR	MARS	RCMARS			
EV	0.860	0.870	0.598	0.748	0.779	0.850	0.862	0.927	0.599	0.740	0.885	0.905
R^2_{adj}	0.831	0.852	0.756	0.832	0.853	0.914	0.835	0.913	0.746	0.836	0.914	0.92
AAE	0.255	0.251	0.277	0.208	0.194	0.156	0.193	0.204	0.169	0.137	0.129	0.128
RMSE	0.367	0.353	0.396	0.328	0.307	0.235	0.363	0.264	0.363	0.292	0.212	0.205
r	0.927	0.933	0.928	0.946	0.952	0.972	0.929	0.963	0.944	0.960	0.977	0.978
	LR	MARS	RCMARS				LR	MARS	RCMARS			
EV	0.935	0.965	0.587	0.719	0.835	0.953	0.847	0.938	0.490	0.569	0.841	0.882
R^2_{adj}	0.922	0.960	0.875	0.923	0.945	0.960	0.816	0.926	0.772	0.817	0.927	0.960
AAE	0.204	0.153	0.253	0.195	0.168	0.149	0.263	0.182	0.264	0.217	0.104	0.086
RMSE	0.250	0.182	0.30	0.236	0.200	0.171	0.383	0.244	0.394	0.353	0.224	0.166
r	0.967	0.983	0.974	0.979	0.981	0.985	0.920	0.969	0.949	0.954	0.976	0.987

Table B.2 Performance measures of LR, MARS and RCMARS models based on different simulated data for each environmental item

e_1							e_2					
	LR	MARS	RCMARS				LR	MARS	RCMARS			
EV	0.839	0.912	0.590	0.664	0.888	0.907	0.848	0.860	0.620	0.667	0.704	0.834
R^2_{adj}	0.809	0.894	0.780	0.810	0.895	0.901	0.818	0.840	0.785	0.819	0.842	0.910
AAE	0.268	0.243	0.268	0.241	0.201	0.196	0.275	0.285	0.241	0.221	0.211	0.166
RMSE	0.391	0.291	0.375	0.349	0.260	0.252	0.382	0.367	0.382	0.351	0.328	0.248
r	0.917	0.955	0.939	0.943	0.965	0.966	0.921	0.927	0.932	0.942	0.949	0.969
	LR	MARS	RCMARS				LR	MARS	RCMARS			
EV	0.805	0.853	0.551	0.634	0.694	0.835	0.564	0.863	0.380	0.558	0.749	0.825
R^2_{adj}	0.765	0.832	0.704	0.781	0.834	0.920	0.478	0.835	0.661	0.779	0.829	0.835
AAE	0.274	0.251	0.335	0.273	0.233	0.173	0.407	0.273	0.404	0.334	0.289	0.277
RMSE	0.434	0.376	0.449	0.386	0.336	0.234	0.646	0.363	0.521	0.421	0.370	0.364
r	0.897	0.924	0.903	0.929	0.946	0.973	0.751	0.929	0.890	0.919	0.928	0.929
	LR	MARS	RCMARS				LR	MARS	RCMARS			
EV	0.691	0.917	0.483	0.662	0.883	0.897	0.846	0.839	0.585	0.755	0.790	0.817
R^2_{adj}	0.629	0.895	0.630	0.800	0.896	0.900	0.815	0.816	0.752	0.818	0.832	0.840
AAE	0.295	0.207	0.277	0.200	0.174	0.171	0.185	0.221	0.250	0.198	0.184	0.179
RMSE	0.545	0.282	0.487	0.358	0.259	0.254	0.385	0.393	0.423	0.362	0.348	0.340
r	0.831	0.958	0.889	0.939	0.965	0.966	0.920	0.916	0.914	0.931	0.936	0.939
	LR	MARS	RCMARS				LR	MARS	RCMARS			
EV	0.702	0.908	0.431	0.586	0.790	0.884	0.874	0.860	0.640	0.772	0.781	0.856
R^2_{adj}	0.640	0.889	0.641	0.777	0.890	0.921	0.850	0.847	0.752	0.848	0.851	0.900
AAE	0.398	0.244	0.361	0.285	0.196	0.159	0.185	0.223	0.233	0.189	0.181	0.162
RMSE	0.536	0.297	0.480	0.378	0.266	0.226	0.347	0.367	0.359	0.282	0.279	0.229
r	0.837	0.953	0.907	0.939	0.966	0.974	0.935	0.927	0.941	0.961	0.962	0.974

(continued)

Table B.2 (continued)

e_1							e_2					
	LR	MARS	RCMARS				LR	MARS	RCMARS			
EV	0.561	0.711	0.411	0.639	0.696	0.754	0.872	0.910	0.656	0.826	0.871	0.908
R^2_{adj}	0.474	0.685	0.530	0.686	0.709	0.720	0.846	0.892	0.750	0.847	0.893	0.910
AAE	0.520	0.446	0.476	0.370	0.354	0.345	0.284	0.241	0.305	0.229	0.189	0.176
RMSE	0.649	0.527	0.565	0.462	0.445	0.437	0.351	0.294	0.361	0.282	0.236	0.217
r	0.749	0.843	0.841	0.886	0.893	0.896	0.934	0.954	0.939	0.959	0.971	0.975

Appendix C
Sensitivity and Performance of MARS for Forecasting of Precipitation

See Tables C.1 and C.2.

Note: *indicates the best performance of ***U*** for train data, test data and stability (st) with respect to the related performance measure.

Table C.1 For sensitivity analysis: parameter values of RCMARS model based on different uncertainty scenarios

U	**±3/5**				**±3/10**			
v	**±3/5**	**±3/10**	**±3/20**	**±0**	**±3/5**	**±3/10**	**±3/20**	**±0**
α_0	−0.707	−0.729	−0.735	−0.788	−0.339	−0.329	−0.353	−0.590
α_1	0.410	0.411	0.408	0.366	0.412	0.403	0.395	0.379
α_2	0.371	0.440	0.480	0.422	0.546	0.581	0.613	0.516
α_3	−0.334	−0.376	−0.391	−0.322	−0.276	−0.271	−0.271	−0.298
α_4	0.390	0.551	0.651	0.571	1.274	1.425	1.545	1.030
α_5	0.132	0.105	0.086	0.086	0.030	0.007	−0.014	0.052
α_6	−0.289	−0.355	−0.386	−0.241	−0.540	−0.555	−0.564	−0.374
α_7	−0.291	−0.322	−0.335	−0.297	−0.350	−0.358	−0.365	−0.329
α_8	−0.163	−0.243	−0.286	−0.200	−0.393	−0.434	−0.471	−0.332
α_9	0.000	0.000	0.000	0.000	0.000	0.000	0.000	−0.001
α_{10}	0.000	0.000	−0.004	0.000	−0.187	−0.203	−0.211	−0.102
α_{11}	−0.240	−0.212	−0.192	−0.128	−0.396	−0.371	−0.336	−0.252
α_{12}	−0.124	−0.112	−0.096	−0.065	−0.100	−0.070	−0.039	−0.066

(continued)

A. Özmen, *Robust Optimization of Spline Models and Complex Regulatory Networks*, Contributions to Management Science,
DOI 10.1007/978-3-319-30800-5

Table C.1 (continued)

U	**±3/20**				**±0**			
v	**±3/5**	**±3/10**	**±3/20**	**±0**	**±3/5**	**±3/10**	**±3/20**	**±0**
α_0	−0.255	−0.256	−0.367	−0.367	−0.187	−0.187	−0.187	0.065
α_1	0.402	0.390	0.386	0.386	0.398	0.398	0.398	0.390
α_2	0.551	0.603	0.558	0.559	0.550	0.550	0.550	0.675
α_3	−0.245	−0.251	−0.268	−0.268	−0.233	−0.233	−0.233	−0.181
α_4	1.363	1.528	1.288	1.288	1.384	1.384	1.385	2.065
α_5	0.034	0.005	0.034	0.034	0.036	0.036	0.036	−0.062
α_6	−0.512	−0.531	−0.453	−0.453	−0.500	−0.500	−0.500	−0.665
α_7	−0.345	−0.360	−0.343	−0.343	−0.342	−0.342	−0.342	−0.377
α_8	−0.402	−0.454	−0.395	−0.395	−0.406	−0.406	−0.406	−0.562
α_9	−0.020	−0.027	−0.029	−0.029	−0.051	−0.051	−0.051	−0.080
α_{10}	−0.219	−0.233	−0.178	−0.178	−0.230	−0.230	−0.230	−0.351
α_{11}	−0.407	−0.359	−0.315	−0.315	−0.390	−0.390	−0.390	−0.386
α_{12}	−0.110	−0.086	0.094	−0.094	−0.134	−0.134	−0.134	−0.104

Table C.2 Performance measures of RCMARS for different uncertainty scenarios

U	±3/5											
v	±3/5			±3/10			±3/20			±0		
	Train	Test	St	Train	Test	St	Train	Test	St	Train	Test	St
R^2	0.782	0.833	0.939	0.813	0.849	0.958	0.823*	0.846	0.972*	0.767	0.850*	0.902
AAE	0.360	0.256	0.710*	0.327	0.206	0.631	0.312*	0.200*	0.641	0.374	0.256	0.683
RMSE	0.459	0.366	0.796	0.426	0.348	0.818	0.414*	0.351	0.848*	0.475	0.347*	0.730
r	0.907	0.942	0.962	0.914	0.935	0.977	0.916*	0.931	0.984*	0.898	0.944*	0.952
EV	0.504	0.523	0.963	0.602	0.669	0.899	0.649	0.741	0.876	0.495	0.585	0.846
U	±3/10											
v	±3/5			±3/10			±3/20			±0		
	Train	Test	St	Train	Test	St	Train	Test	St	Train	Test	St
R^2	0.934	0.649	0.695	0.941	0.609	0.647	0.943*	0.588	0.623	0.876	0.789*	0.901*
AAE	0.198	0.410	0.482	0.181	0.441	0.410	0.172*	0.462	0.372	0.273	0.311*	0.877*
RMSE	0.253	0.530	0.477	0.238	0.559	0.426	0.234*	0.574	0.408	0.346	0.411*	0.842*
r	0.973	0.823	0.846	0.975	0.811	0.831	0.975*	0.808	0.828	0.950	0.900*	0.947*
EV	0.752	0.800	0.940*	0.788	0.880	0.894	0.820	0.954	0.859	0.628*	0.687*	0.914

(continued)

Table C.2 (continued)

U	**±3/20**											
v	**±3/5**			**±3/10**			**±3/20**			**±0**		
	Train	Test	St	Train	Test	St	Train	Test	St	Train	Test	St
R^2	0.934	0.649	0.695	0.941	0.609	0.647	0.943*	0.588	0.623	0.876	0.789*	0.901*
AAE	0.198	0.410	0.482	0.181	0.441	0.410	0.172*	0.462	0.372	0.273	0.311*	0.877*
RMSE	0.253	0.530	0.477	0.238	0.559	0.426	0.234*	0.574	0.408	0.346	0.411*	0.842*
r	0.973	0.823	0.846	0.975*	0.811	0.831	0.975	0.808	0.828	0.950	0.900*	0.947*
EV	0.752	0.800	0.940*	0.788	0.880	0.894	0.820	0.954	0.859	0.628*	0.687*	0.914
U	**±0**											
v	**±3/5**			**±3/10**			**±3/20**			**±0**		
	Train	Test	St	Train	Test	St	Train	Test	St	Train	Test	St
R^2	0.941	0.563	0.598	0.941	0.563*	0.598	0.941	0.563	0.598*	0.971*	0.225	0.231
AAE	0.186	0.468	0.398	0.186	0.468	0.398	0.186	0.468*	0.398*	0.131*	0.646	0.203
RMSE	0.239	0.591	0.404	0.239	0.591	0.404	0.239	0.591*	0.404*	0.166*	0.788	0.211
r	0.977	0.769	0.787*	0.977	0.769	0.787*	0.977	0.769*	0.787*	0.986*	0.672	0.682
EV	0.735*	0.788*	0.932*	0.735*	0.788	0.932*	0.735*	0.789	0.931	0.953	1.241	0.768

Appendix D
Prediction Performance Criteria and Related Measures

Notes:

N: number of observations;
p: number of terms in the model;
y_k: kth observed response value;
$\hat{y}_k$: kth estimated (fitted) response value;
$\bar{y}$: mean of the observed values;
$\hat{y}$: estimated response variable;
$\bar{\hat{y}}$: mean of the estimated response variable;
$s(y)^2$: standard deviation of the observed response variable;
$s(\hat{y})^2$: standard deviation of the estimated response variable;
M_{TR} and M_{TE}: the measure values for training and test data, respectively.

See Table D.1.

A. Özmen, *Robust Optimization of Spline Models and Complex Regulatory Networks*, Contributions to Management Science,
DOI 10.1007/978-3-319-30800-5

Table D.1 Prediction performance criteria and related measures

Criterion	Abbreviation	Measure (M)	Explanation	Interpretation	Formula
Accuracy	R^2	Multiple Coefficient of Determination	Percentage of variation in response explained by the model	Values closer to one are better	$R^2 := 1 - \left(\frac{\sum_{k=1}^{N}(y_k-\hat{y}_k)^2}{\sum_{k=1}^{n}(y_k-\bar{y}_k)^2}\right).$
	R^2_{adj}	Adjusted Multiple Coefficient of Determination	Percentage of variation in response explained by the model	Values closer to one are better	$R^2_{adj} := 1 - \left(\frac{\sum_{k=1}^{N}(y_k-\hat{y}_k)^2}{\sum_{k=1}^{n}(y_k-\bar{y}_k)^2}\right) \cdot \left(\frac{N-1}{N-p-1}\right).$
	AAE	Average Absolute Error	Average magnitude of errors	Smaller values are better	$AAE := \frac{1}{N}\sum_{k=1}^{N} \lvert y_k - \hat{y}_k \rvert.$
	RMSE	Root Mean Square Error	Average magnitude of errors	Smaller values are better	$RMSE := \sqrt{\frac{1}{N}\sum_{k=1}^{N}(y_k - \hat{y}_k)^2}.$
	r	Correlation coefficient	Linear relation between observed and predicted response	Values closer to one are better	$r := \frac{\sum_{k=1}^{N}(y-\bar{y})(\hat{y}-\bar{\hat{y}})/(N-1)}{\sqrt{s(y)^2 s(\hat{y})^2}}.$
Precision	*EV*	Estimation Variance	Variance of the estimated response values	Smaller values are better	$EV := \frac{\sum_{k=1}^{N}(\hat{y}_k-\bar{\hat{y}})^2}{N-1}.$
Stability	–	Stability of a measure	Compares the performance of a method on both training and test data	Values closer to one indicate more stable models	$\mathrm{Min}\{\frac{M_{TR}}{M_{TE}}, \frac{M_{TE}}{M_{TR}}\}.$

References

1. A. Abraham and D. Steinberg, Is neural network a reliable forecaster on earth? A MARS query!, Bio-Inspired Applications of Connectionism, 2085, pp. 679–686, 2001.
2. A. Abraham, D. Steinberg and N.S. Philip, Rainfall forecasting using soft computing models and multivariate adaptive regression splines, IEEE SMC Transactions, 1, pp. 1–6, 2001.
3. K. D. Andersen, *Minimizing a Sum of Norms (Large Scale solutions of symmetric positive definite linear systems)*, PhD thesis, Odense University, 1995.
4. E. D. Andersen, C. Roos and T. Terlaky, On implementing a primal-dual interior-point method for conic quadratic optimization, Mathematical Programming, Ser. B 95, pp. 249–277, 2003.
5. R. C. Aster, B. Borchers and C. Thurber, *Parameter Estimation and Inverse Problems*, Academic Press, 2004.
6. I. Batmaz and G. Köksal, Overview of knowledge discovery in databases process and data mining for surveillance technologies and EWS. In Surveillance Technologies and Early Warning Systems: Data Mining Applications for Risk Detection, A.S. Koyuncugil and N. Ozgulbas (Eds.), Hershey, PA: IGI Global Publisher (Idea Group Publisher), pp. 1–30, 2011.
7. P.A. Bekker, Comment on identification in the linear errors in variables model, Econometrica, 54 (1), pp. 215–217, 1986.
8. A. Ben-Tal and A. Nemirovski, Robust truss topology design via semidefinite programming, SIAM Journal on Optimization, 7(4), pp. 991–1016, 1997.
9. A. Ben-Tal and A. Nemirovski, Robust convex optimization, Mathematics of Operations Research, 23, pp. 769–805, 1998.
10. A. Ben-Tal and A. Nemirovski, Robust solutions to uncertain linear programs, Operations Research Letters, 25(1), pp. 1–13, 1999.
11. A. Ben-Tal and A. Nemirovski, Robust solutions of linear programming problems contaminated with uncertain data, Mathematical Programming, 88, pp. 411–424, 2000.
12. A. Ben-Tal and A. Nemirovski, Lectures on Modern Convex Optimization: Analysis, Algorithms, and Engineering Applications, MPR-SIAM Series on Optimization, SIAM, Philadelphia, 2001.
13. A. Ben-Tal, L. El-Ghaoui and A. Nemirovski, *Robust semidefinite programming*, R. Saigal, H. Wolkowitcz, L. Vandenberghe, Eds. Handbook on Semidefinite programming and applications, KluwerAcademic Publishers, pp. 139–162, 2000.
14. A. Ben-Tal and A. Nemirovski, Robust optimization - methodology and applications, Mathematical Programming, 92(3), pp. 453–480, 2002.
15. A. Ben-Tal, L. El-Ghaoui, and A. Nemirovski, *Robust Optimization*, Princeton University Press, 2009.

A. Özmen, *Robust Optimization of Spline Models and Complex Regulatory Networks*, Contributions to Management Science,
DOI 10.1007/978-3-319-30800-5

16. V.J. Beck and K. J. Arnold, *Parameter Estimation in Engineering and Science*, John Wiley and Sons, 1977.
17. D.P. Bertsekas, *Dynamic Programming and Optimal Control*, Athena Scientific, Belmont, Mass., 1995.
18. D. Bertsimas, M. Sim, Price of Robustness, Oper. Res. 52(1), pp. 35–53, 2004.
19. D. Bertsimas and M. Sim, Tractable Approximations to Robust Conic Optimization Problems, Mathematical Programming, Ser. B, 107, pp. 5–36, 2006.
20. D. Bertsimas, D.B. Brown, C. Caramanis, Theory and applications of Robust Optimization, Technical Report, University of Texas, Austin, TX, USA, 2007.
21. O. Boni, *Robust Solutions of Conic Quadratic Problems*, PhD Thesis, Technion, Israeli Institute of Technology, IEM faculty, 2007.
22. B. Bower. Banks err by confusing risk, uncertainty, Science News, 182(10), pp. 13, 2012.
23. E. Borenstein and M.W. Feldman, Topological signatures of species interactions in metabolic networks, Journal of Computational Biology, 16(2), pp.191–200, 2009.
24. L. Breiman, J. Friedman, R. Olshen and C. Stone, *Classification and Regression Trees*, Belmont, CA: Wadsworth Int. Group, 1984.
25. Central Bank of the Republic of Turkey: http://www.tcmb.gov.tr.
26. A. Charnes and W.W. Cooper, Chance constrained programming, Management Science, 6, pp. 73–89, 1959.
27. T. Chen, H. L. He and G. M. Church, Modeling gene expression with differential equations, Pacific Symposium on Biocomputing, 4, pp. 29–40, 1999.
28. A. Chesher, The effect of measurement error, Biometrika, 78(3), pp. 451–462, 1991.
29. J. Corte-Real, X. Zhang and X. Wang, Downscaling GCM information to regional scale: a non- parametric multivariate regression approach, Climate Dynamics, 11, pp. 413–424, 1995.
30. Z. Çavuşoğlu, *Predicting Debt Crises in Emerging Markets Using Generalized Partial Linear Models*, Term Project, Institute of Applied Mathematics, Middle East Technical University, Ankara, 2010.
31. G. Çelik, *Parameter Estimation in Generalized Partial Linear Models with Conic Quadratic Programming*, M.Sc. Thesis, Institute of Applied Mathematics, METU, Ankara, 2010.
32. K. Deb, *Multi Objective Optimization using Evolutionary Algorithms*, John Wiley and Sons, 2001.
33. E. Detragiache and A. Spilimbergo, Short-Term Debt and Crises, International Money Fund. European Summer Symposium in International Macroeconomics, Israel, 2001.
34. B. Efron, R. Tibshirani, *An Introduction to the Bootstrap*, Boca Raton, FL: Chapman and Hall/CRC, 1993.
35. L. El-Ghaoui, Robust Optimization and Applications, IMA Tutorial, 2003.
36. L. El-Ghaoui and H. Lebret, Robust solutions to least-square problems to uncertain data matrices, SIAM Journal on Matrix Analysis and Applications, 18, pp. 1035–1064, 1997.
37. L. El-Ghaoui, F. Oustry and H. Lebret, Robust solutions to uncertain semidefinite programs, SIAM Journal on Optimization 9, pp. 33–52, 1998.
38. I. Elishakoff, Whys and Hows in Uncertainty Modelling, Probability, Fuzziness and Anti-Optimization, 1999.
39. S.V. Emmerik, Risk, uncertainty and the financial crisis – Reflexivity in Finance, January 20, 2009. Page available at: http://reflexivityfinance.blogspot.com/2009/01/risk-uncertainty-and-financial-crisis.html.
40. F.J. Fabozzi, P.N. Kolm, D.A. Pachamanova and S.M. Focardi, *Robust Portfolio Optimization and Management*, Willey Finance, 2007.
41. J.E. Falk, Exact solutions of inexact linear programs, Operations Research, 24, pp. 783–787, 1976.
42. J.A. Filar and A. Haurie, *Uncertainty and Enviromental Decision Making*, Springer, 2010.
43. M. Fioramanti, Predicting Sovereign Debt Crises Using Artificial Neural Networks: A Comparative Approach, Journal of Financial Stability, 4(2), pp. 149–164, 2008.
44. J. Fox, *Bootstrapping Regression Models: An R and S-PLUS Companion to Applied Regression*, Sage Publications, CA, USA, 2002.

45. J. Fox, *Nonparametric Regression*, in: B. Everitt and D. Howell eds., Encyclopedia of Statistics in the Behavioral Sciences, London: Wiley, 2005.
46. J.H. Friedman, Multivariate adaptive regression splines, The Annals of Statistics, 19(1), pp. 1–141, 1991.
47. J. Gebert, M. Lätsch, S.W. Pickl, G.-W. Weber and R. Wünschiers: An algorithm to analyze stability of gene-expression pattern, In M. Anthony, E. Boros, P.L. Hammer, and A. Kogan (guest eds.), special issue *Discrete Mathematics and Data Mining II* of Discrete Applied Mathematics, 154(7), pp. 1140–1156, 2006.
48. G. Gigerenzer, Cognitive foundations of risk judgments, 7th DFG-NSF conference - Reckoning with the risk of catastrophe, Washington, D.C., October 4, 2012, abstract available at: http://dfg-nsf2012.mit.edu/wp-content/uploads/2012/09/Conference-brochure.pdf.
49. A. Gökmen, S. Kayalgil, G.-W. Weber, I. Gökmen, M. Ecevit, A. Sürmeli, T. Bali, Y. Ecevit, H. Gökmen and D.J. DeTombe, Balaban Valley Project: Improving the quality of life in rural area in Turkey. International Scientific Journal of Methods and Models Complexity, 7(1), 2004.
50. A.M. Gonzalez, A.M.S. Roque and J. Garcia-Gonzalez, Modeling and forecasting electricity prices with input/output hidden Markov models. IEEE Transaction on Power Systems, 20, pp. 13–24, 2005.
51. D.N. Gujarati and D.C. Porter, Basic Econometrics, McGraw-Hill, Boston, 2009.
52. O. Güler, Barrier functions in interior-point methods, Mathematics of Operations Research, 21, pp. 860–885, 1996.
53. J. Hadamard, *on Cauchy's Problem in Linear Partial Differential Equations*, Yale University Press, New Haven, 1923.
54. P.C. Hansen and D. P. O'Leary, The use of the L-curve in the regularization of discrete ill-posed problems, SIAM Journal on Scientific Computing, 14(6), pp. 1487–1503, 1993.
55. P.C. Hansen, Regularization tools: A Matlab package for analysis and solution of discrete ill-posed problems, Numerical Algorithms, 6(I-II), pp. 1–35, 1994.
56. J.R. Harris, W. Nystad and P. Magnus, Using genes and environments to define asthma and related phenotypes: applications to multivariate data. Clinical and Experimental Allergy, 28(1), pp. 43–45, 1998.
57. T.J. Hastie and R.J. Tibshirani, *Generalized Additive Models*, Chapman and Hall, London, 1990.
58. T. Hastie, R. Tibshirani and J. H. Friedman, *The Element of Statistical Learning*, Springer Verlag, New York, 2001.
59. M.D. Hoon, S. Imoto, K. Kobayashi, N. Ogasawara and S. Miyano, Inferring gene regulatory networks from time-ordered gene expression data of Bacillus Subtilis using differential equations, Pacific Symposium on Biocomputing, 8, pp. 17–28, 2003.
60. C. İyigün, M. Türkeş, İ. Batmaz, C. Yozgatlıgil, V. Purutcuoglu, E. Kartal-Koç and M. Z. Öztürk, Clustering current climate regions of Turkey by using a multivariate statistical method, Theoretical and Applied Climatology, 114 (1–2), pp. 95–106, 2013.
61. H.D. Jong, Modeling and simulation of genetic regulatory systems: a literature review, Journal of Computational Biology, 9, pp. 103–129, 2002.
62. N. Karamakar, A new polynomial-time algorithm for linear programming, Combinatorica, 4, pp. 373–395, 1984.
63. B. Kayhan, *Parameter Estimation in Generalized Partial Linear Models with Tikhonov Regularization Method*, M.Sc. Thesis, Institute of Applied Mathematics, METU, Ankara, 2010.
64. A. Kibzun, Y. Kan, *Stochastic Programming Problems with Probability and Quantile Functions*, Wiley, 1996.
65. M. Kojima, S. Mizuno and A. Yoshise, A primal-dual interior point algorithm for linear programming, In N. Megiddo, editor, Progress in Mathematical Programming: Interior-Point Algorithms and Related Methods, pp. 29–47, Springer Verlag, Berlin, 1989.
66. W. Krabs and S. Pickl, A game-theoretic treatment of a time-discrete emission reduction model, International Game Theory Review, (6)1, pp. 21–34, 2004.

67. D. Krawczyk-Stando and M. Rudnicki, Regularization Parameter Selection In Discrete Ill-Posed Problems -The Use Of The U-Curve, International Journal of Applied Mathematics and Computer Science, 17(2), pp. 157–164, 2007.
68. M. Kriner, Survival Analysis with Multivariate adaptive Regression Splines, Dissertation, LMU Munchen: Faculty of Mathematics, Computer Science and Statistics, 2007.
69. E. Kropat, S. Pickl and A.Rössler, G. W. Weber, On theoretical and practical relations between discrete optimization and nonlinear optimization, Special issue Colloquy Optimization Structure and Stability of Dynamical Systems (at the occasion of the colloquy with the same name, Cologne, October 2000) of Journal of Computational Technologies, 7, pp. 27–62, 2002.
70. E. Kropat and G.-W. Weber, Robust regression analysis for gene-environment and eco-finance networks under polyhedral and ellipsoidal uncertainty, preprint 2 at Institute of Applied Mathematics, METU, submitted to Optimization Methods and Software, 2010.
71. E. Kropat, G.-W. Weber and J.-J. Rückmann, Regression analysis for clusters in gene-environment networks based on ellipsoidal calculus and optimization, in the special issue in honour of Professor Alexander Rubinov of Dynamics of Continuous, Discrete and Impulsive Systems, Series B: Applications & Algorithms **17**, 5, pp. 639–657, 2010.
72. E. Kropat, G.-W. Weber and B. Akteke-Öztürk, Eco-Finance networks under uncertainty. Proceedings of the International Conference on Engineering Optimization (CD), Engineering Optimization Rio de Janeiro, Brazil, ISBN 978857650156-5, 2008.
73. E. Kropat, G.-W. Weber and C.S. Pedamallu, Regulatory networks under ellipsoidal uncertainty – Data Analysis and Prediction by Optimization theory and dynamical systems, in the book on Data Mining: Foundations and Intelligent Paradigms, 24, pp. 27–56, Springer-Verlag, Berlin, Heidelberg, 2012.
74. R.J. Kuligowski and A.P. Barros, Localized precipitation forecasts from a numerical weather prediction model using artificial neural networks, Weather and Forecasting, 13(4), pp. 1194–1204, 1998.
75. Kyoto (1997), Kyoto Contract, Page available at http://www.unfccc.org/resource/convkp.html
76. G. Lee, T.K. Sung and N. Chang, Dynamics of Modeling in Data Mining: Interpretive Approach to Bankruptcy Prediction, Journal of Management Information Systems, 16, pp. 63–85, 1999.
77. S. Lee, S. Cho and P.M. Wong, Rainfall prediction using artificial neural networks, Journal of Geographic Information and Decision Analysis, 2 (2), pp. 233–242, 1998.
78. Y.F. Li, S. Venkatesh and D. Li, Modeling global emissions and residues of pesticides, Environmental Modeling and Assessment, 9, pp. 237–243, 2004.
79. M.S. Lobo, L. Vanderberghe, S. Boyd and H. Lebret, Applications of second-order cone programming, Linear Algebra and its Applications, pp. 193–228, 1998.
80. J.Löfberg, YALMIP: A Toolbox for Modeling and Optimization in MATLAB, 2004. http://users.isy.liu.se/johanl/yalmip.php.
81. I.J. Lustig, R.E. Marsten, D.F. Shanno, Interior point methods for linear programming: Computational state of the art, ORSA Journal on Computing, 6(1), pp. 1–15, 1994.
82. P. Manasse, N. Roubini and A. Schimmelpfennig, Predicting Sovereign Debt Crises, IMF Working Paper 03/221, International Monetary Fund, 2003, ISBN: 978-1-45187-525-6.
83. MARS Salford Systems, software available at http://www.salfordsystems.com.
84. Minitab, software available at http://www.minitab.com.
85. D.C. Montgomery and G.C. Runger, *Applied Statistics and Probability for Engineers*, New York: John Wiley and Sons, 2007.
86. D.C. Montgomery, *Design and Analysis of Experiment*, Seventh Edition, Wiley, 2009.
87. MOSEK, A very powerful commercial software for CQP, http://www.mosek.com (accessed 05 Sep. 2008).
88. R.D.C. Monteiro and I. Adler, Interior path following primal-dual algorithms. Part I: Linear programming, Mathematical Programming, 44, pp. 27–41, 1989.
89. R.D.C. Monteiro and T. Tsuchiya, Polynomial convergence of primal-dual algorithms for the second order cone program based on the MZ-family of directions, Mathematical Programming, 88(1), pp. 61–83, 2000.

90. M. Müller, Estimation and Testing in Generalized Partial Linear Models – A Comparive Study, Statistics and Computing, 11, pp. 299–309, 2001.
91. M.T. Nair, M. Hegland and R. S. Anderssen, The Trade-off between Regularity and Stability in Tikhonov Regularization, Mathematics of Computation, 66, 217, pp. 193–206, 1997.
92. Y.E. Nesterov and A. Nemirovski, *Interior Point Methods in Convex Programming*, SIAM, 1993.
93. Y. Nesterov and M. J. Todd, Self-scaled barriers and interior-point methods for convex programming, Mathematics of Operations Research, 22(1), pp. 1–42, 1997.
94. F.J. Nogales, J. Contreras, A. J. Conejo and R. Espinola, Forecasting next-day electricity prices by time series models, IEEE Transactions of Power Systems, 17, pp. 342–348, 2002.
95. B.W. Otok, Development of rainfall forecasting model in indonesia by using ASTAR, transfer function, and ARIMA methods, European Journal of Scientific Research, 38(3), pp. 386–395, 2009.
96. A. Özmen, *Robust Conic Quadratic Programming Applied to Quality Improvement- A Robustification of CMARS*, Ms. Thesis, METU, Ankara, Turkey, 2010.
97. A. Özmen, G.-W. Weber and İ. Batmaz, The new robust CMARS (RCMARS) method, in ISI Proceedings of 24th MEC-EurOPT 2010 – Continuous Optimization and Information-Based Technologies in the Financial Sector, İzmir, Turkey, pp. 362–368, 2010.
98. A. Özmen, G.-W. Weber, İ. Batmaz, and E. Kropat, RCMARS: Robustification of CMARS with Different Scenarios under Polyhedral Uncertainty Set, Communications in Nonlinear Science and Numerical Simulation, 16 (12), pp. 4780–4787, 2011.
99. A. Özmen and G.-W. Weber, Robust conic generalized partial linear models using RCMARS method, AIP Conference Proceeding 1499, pp. 337–343, 2012.
100. A. Özmen, G.-W. Weber and E. Kropat, Robustification of conic generalized partial linear models under polyhedral uncertainty, International IFNA-ANS scientific Journal "Problems of Nonlinear Analysis in Engineering Systems", 2(38), 18, pp. 104–113, 2012.
101. A. Özmen, G.-W. Weber and A. Karimov, A new robust optimization tool applied on financial data, to appear in Pacific Journal of Optimization, 9(3), pp. 535–552, 2013.
102. A. Özmen, G.-W. Weber, Z. Çavuşoğlu and Ö. Defterli, The new robust conic GPLM method with an application to finance: prediction of credit default, Journal of Global Optimization, 56(2), pp. 233–249, 2013.
103. A. Özmen, E. Kropat and G.-W. Weber, Spline Regression Models for Complex Multi-modal Regulatory Networks, Optimization Methods and Software, 29(3), pp. 515–534, 2014.
104. A. Özmen, İ. Batmaz and G.-W. Weber, Precipitation Modeling by Polyhedral RCMARS and Comparison with MARS and CMARS, Environmental Modeling and Assessment, 19(5), pp. 425–435, 2014.
105. A. Özmen and G.-W. Weber, RMARS: Robustication of Multivariate Adaptive Regression Spline, and an application in finance, Journal of Computational and Applied Mathematics, 259, pp. 914–924, 2014.
106. A. Özmen, E. Kropat and G.-W. Weber, Robust Optimization in Spline Regression Models for Multi-model Regulatory Networks under Polyhedral Uncertainty, preprint at IAM of METU, submitted to Optimization.
107. T. Partal and H.K. Cigizoglu, Prediction of daily precipitation using wavelet—neural networks, Hydrological sciences journal, 54(2), pp. 234–246,2009.
108. M. Partner, N. Kashtan and U. Alon, Environmental variability and modularity of bacterial metabolic network, BMC Evolutionary Biology 7(169), 2007.
109. S. Pickl and G.-W. Weber, Optimization of a time-discrete nonlinear dynamical system from a problem of ecology - an analytical and numerical approach, Journal of Computational Technologies, 6(1), pp. 43–52, 2001.
110. S. Pickl, An iterative solution to the nonlinear time-discrete TEM model - the occurence of chaos and a control theoretic algorithmic approach, AIP Conference Proceedings, 627(1), pp. 196–205, 2002.
111. I. Popescu, Robust mean-covariance solutions for stochastic optimization, Operations Research, 55 (1), 98–112, 2007.

112. Prajneshu, Cautionary note About Nonlinear Models in Fisheries, Indian Journal of Fisheries, 38, pp. 231–33, 1991.
113. Prajneshu, A Nonlinear statistical Model for Aphid Population Growth, Journal of the Indian Society of Agricultural Statistics, 51, pp. 73–80, 1998.
114. C. Roos, T. Terlaky and J. Vial, *Interior point approach to linear optimization: theory and algorithms*, John Wiley and Sons, New York, 1997.
115. C. Roos, T. Terlaky and J. Vial, *Interior Point Methods for Linear Optimization*, Springer Science, Heidelberg/Boston, 2006.
116. R.Y. Rubinstein and D.P. Kroese, *Simulation and the Monte Carlo Method*, (2nd ed.), New York: John Wiley and Sons, 2007.
117. E. Sakamoto and H. Iba, Inferring a system of differential equations for a gene regulatory network by using genetic programming, Proceeding of Congress on Evolutionary Computing 01, pp. 720–726, 2001.
118. S.H. Schmieta and F. Alizadeh, *Associative algebras, symmetric cones and polynomial time interior point algorithms*, Technical Report RRR 17–98, RUTCOR, Rutgers Center for Operations Research, P.O. Box 5062, New Brunswick, New Jersey, 1998.
119. G.F. Seber and C. J. Wild, *Nonlinear Regression*, John Wiley and Sons, 1989.
120. M. Sim, *Robust Optimization*, PhD Thesis, Massachusetts Institute of Technology, Cambridge MA, 2004.
121. A.L. Soyster, Convex programming with set-inclusive constraints and applications to inexact linear programming, Operations Research, 21, pp. 1154–1157, 1973.
122. Copyright StatSoft, Inc. Multivariate Adaptive Regression Splines, http://www.statsoft.com/textbook/stmars.html (accessed 05 Sep. 2008).
123. R.E. Steuer, *Multiple Criteria Optimisation: Theory, Computation and Application*, New York: John Wiley and Sons, NY, 1986.
124. J.F. Sturm, Primal-Dual Interior Point Approach to Semidefinite Programming, Vol.156 of Tinbergen Institute Research Series, Thesis Publishers, The Netherlands, 1997.
125. J. Sturm, Using SeDuMi 1.02 a MATLAB Toolbox for Optimization Over Symmetric Cones, Optimization Methods and Software Research, 11–12, pp. 625–653, 1999.
126. A. Tarantola,*Inverse Problem Theory and Methods for Model Parameter Estimation*, SIAM, 2005.
127. P. Taylan, G.-W. Weber and A. Beck, New approaches to regression by generalized additive models and continuous optimization for modern applications in finance, science and technology, Journal Optimization 56, 5–6, pp. 1–24, 2007.
128. M. Tranmer and M. Elliot, Binary logistic regression, Cathie Marsh for Census and Survey Research, Paper 20, 2008.
129. T. Tsuchiya, *A polynomial primal-dual path-following algorithm for second-order cone programming*, Technical report, The Institute of Statistical Mathematics, Tokyo, Japan, 1997.
130. M. Türkeş, *Klimatoloji and meteoroloji*, İstanbul, Turkey: Kriter Yayinevi, 2010.
131. R. H. Tütüncü, K.C. Toh and M.J. Todd, Solving semidefinite-quadratic-linear programs using SDPT3, Mathematical Programming Series B, 95, pp. 189–217, 2003.
132. O. Uğur, S. W. Pickl, G.-W. Weber and R. Wünschiers, An algorithmic approach to analyze genetic networks and biological energy production: an introduction and contribution where OR meets biology, Optimization, 58(1), pp. 1–22, 2009.
133. O. Uğur and G.-W. Weber, Optimization and dynamics of gene-environment networks with intervals. Journal of Industrial Management and Optimization 3(2), pp. 357–379, 2007.
134. C. Venkatesan, S.D. Raskar, S.S. Tambe, B.D. Kulkarni and R.N. Keshavamurty, Prediction of all summer monsoon rainfall using error- back-propagation Neural Network, Meterology and Atmospheric Physics, 62, pp. 225–240, 1997.
135. G.-W. Weber, S. Z. Alparslan-Gök and B. Söyler, A new mathematical approach in environmental and life sciences: gene-environment networks and their dynamics, Environmental Modeling and Assesment, 14(2), pp. 267–288, 2007.

136. G.-W. Weber, İ. Batmaz, G. Köksal, P. Taylan and F. Yerlikaya, CMARS: a new contribution to nonparametric regression with multivariate adaptive regression splines supported by continuous optimization, IPSE 20 (3), pp. 371–400, 2012.
137. G.-W. Weber, Z. Çavuşoğlu and A. Özmen, Predicting Default Probabilities in Emerging Markets by New Conic Generalized Partial Linear Models and Their Optimization, Optimization, 61(4), pp. 443–457, 2012.
138. G.-W. Weber, A. Tezel, P. Taylan, A. Söyler and M. Çetin, Mathematical contributions to dynamics and optimization of gene-environment networks, Optimization, 57(2), pp. 353–377, 2008.
139. G.-W. Weber, S.Z. Alparslan-Gök and N. Dikmen, Environmental and life sciences: gene-environment networks - optimization, games and control - a survey on recent achievements, Journal of Organisational Transformation and Social Change, 5(3), pp. 197–233, 2008.
140. G.-W. Weber, P. Taylan, S.-Z. Alparslan-Gök, S. Özöğür and B. Akteke-Öztürk, Optimization of gene-environment networks in the presence of errors and uncertainty with Chebychev approximation. TOP, the Operational Research journal of SEIO (Spanish Statistics and Operations Research Society) 16(2), pp. 284–318, 2008.
141. G.-W. Weber, S. Özögür-Akyüz and E. Kropat, A review on data mining and continuous optimization applications in computational biology and medicine. Embryo Today, Birth Defects Research (Part C) 87, pp. 165–181, 2009.
142. G.-W. Weber, O. Uğur, P. Taylan and A. Tezel, On optimization, dynamics and uncertainty: a tutorial for gene-environment networks, Discrete Applied Mathematics, 157(10), pp. 2494–2513, 2009a.
143. G.-W. Weber, S. Akyüz-Özöğür and E. Kropat, A review on data mining and continuous optimization applications in computational biology and medicine, Birth Defects Research (Part C)-Embryo Today, 87(2), pp. 165–181, 2009b.
144. G.-W. Weber, E. Kropat, A. Tezel, and S. Belen, Optimization applied on regulatory and eco-finance networks - survey and new developments, Pacific Journal of Optimization, 6(2), pp. 319–340, 2010.
145. G.-W. Weber, E. Kropat, B. Akteke-Öztürk and Z.-K. Görgülü, A survey on OR and mathematical methods applied on gene-environment networks, Central European Journal of Operations Research (CEJOR), 17, pp. 315–341, 2009.
146. R. Werner, Cascading: an adjusted exchange method for robust conic programming, CEJOR, 16, pp.179–189, 2008.
147. M. H. Yıldırım, A. Özmen, Ö. Türker Bayrak and G.-W. Weber, Electricity price modeling for Turkey, Operations Research Proceedings 2011, Selected Papers of the International Conference on Operations Research (OR 2011), August 30 - September 2, 2011, Zurich, Switzerland, D. Klatte, K. Schmedders and Hans-Jakob Luethi, eds., pp. 39–44, 2012.
148. C. Yozgatlıgil, S. Aslan, C. İyigün and İ. Batmaz, Comparison of missing value imputation methods for Turkish meteorological time series data, Theoretical and Applied Climatology, 112 (1–2), pp. 143–167, 2012.

MIX
Papier aus verantwortungsvollen Quellen
Paper from responsible sources
FSC® C105338

If you have any concerns about our products,
you can contact us on
ProductSafety@springernature.com

In case Publisher is established outside the EU,
the EU authorized representative is:
Springer Nature Customer Service Center GmbH
Europaplatz 3, 69115 Heidelberg, Germany

Printed by Libri Plureos GmbH
in Hamburg, Germany